Concepts of
Distributed
System

Concepts of
Distributed
System

Manish Varshney PhD
Professor and Head
Department of Computer Science and Engineering
Shri Siddhi Vinayak Institute of Technology
Bareilly, UP

Shanoo Agarwal M Tech
Assistant Professor
Department of Computer Science and Engineering
Shri Siddhi Vinayak Institute of Technology
Bareilly, UP

CBS

CBS Publishers & Distributors Pvt Ltd

New Delhi • Bengaluru • Chennai • Kochi • Kolkata • Mumbai
Hyderabad • Nagpur • Patna • Pune • Vijayawada

Concepts
of
Distributed System

ISBN: 978-81-239-2966-8

Copyright © Authors and Publisher

First Edition: 2016

Published by Satish Kumar Jain and produced by Varun Jain for

CBS Publishers & Distributors Pvt Ltd

4819/XI Prahlad Street, 24 Ansari Road, Daryaganj, New Delhi 110 002, India.

Ph: 23289259, 23266861, 23266867   Website: www.cbspd.com

Fax: 011-23243014   e-mail: delhi@cbspd.com; cbspubs@airtelmail.in.

Corporate Office: 204 FIE, Industrial Area, Patparganj, Delhi 110 092

Ph: 4934 4934   Fax: 4934 4935   e-mail: publishing@cbspd.com; publicity@cbspd.com

Branches

- **Bengaluru:** Seema House 2975, 17th Cross, K.R. Road,
  Banasankari 2nd Stage, Bengaluru 560 070, Karnataka
  Ph: +91-80-26771678/79   Fax: +91-80-26771680   e-mail: bangalore@cbspd.com
- **Chennai:** 7, Subbaraya Street, Shenoy Nagar, Chennai 600 030, Tamil Nadu
  Ph: +91-44-26680620, 26681266   Fax: +91-44-42032115   e-mail: chennai@cbspd.com
- **Kochi:** Ashana House, 39/1904, AM Thomas Road, Valanjambalam,
  Ernakulam 682 018, Kochi, Kerala
  Ph: +91-484-4059061-62-64-65   Fax: +91-484-4059065   e-mail: kochi@cbspd.com
- **Kolkata:** 6/B, Ground Floor, Rameswar Shaw Road, Kolkata-700 014, West Bengal
  Ph: +91-33-22891126, 22891127, 22891128   e-mail: kolkata@cbspd.com
- **Mumbai:** 83-C, Dr E Moses Road, Worli, Mumbai-400018, Maharashtra
  Ph: +91-22-24902340/41   Fax: +91-22-24902342   e-mail: mumbai@cbspd.com

Representatives

- **Hyderabad** 0-9885175004   • **Nagpur** 0-9021734563   • **Patna** 0-9334159340
- **Pune** 0-9623451994   • **Vijayawada** 0-9000660880

Printed at : India Binding House, Noida

$$\frac{to}{God}$$

Preface

A distributed system consists of multiple autonomous computers that communicate through a computer network and its importance is well known in various engineering fields. The computers interact with each other in order to achieve a common goal. This book is structured to cover the key aspects of the subject *concepts of distributed system*.

The book uses simple and lucid language to explain fundamentals of the subject, logical methods of explaining various complicated concepts, and stepwise methods to explain the important topics. Utmost care has been taken to make students comfortable in understanding the basic concepts of the subject. The intended readers are all undergraduate students, scholars, practitioners and technicians seeking an extensive course in the field of distributed systems.

The book is divided into 12 chapters: Chapter 1 provides information on various definitions of distributed system, its requirement and advantages; Chapter 2 covers various system models of distributed computing; Chapter 3 explains the use of logical clocks and vector clocks in distributed system; Chapter 4 illustrates the classification of various distributed mutual algorithms, it also covers various token and nontoken-based algorithms; Chapter 5 covers the distributed deadlock techniques of prevention, avoidance and removal, it also covers the classification of various distributed algorithms like centralized, distributed and hierarchical deadlock detection; Chapter 6 deals with various agreement problems and illustrates the solution of Byzantine agreement problem and its applications; Chapter 7 deals with concepts of DFS, their mechanism and various design issues in its implementation, it covers the logic of DSM, its architecture and various algorithms to implement it; Chapter 8 focuses the failures and its recovery in concurrent systems along with synchronous checkpointing and recovery; Chapter 9 emphasizes on various issues related to fault tolerance, it covers the concept of commit protocols and also highlights the working of dynamic voting based protocol; Chapter 10 covers all the basic concepts of transaction, also covers concepts of locking techniques and various timestamp ordering techniques; Chapter 11 covers various atomic commit protocols; Chapter 12 describes various issues related to replication in distributed computing.

The book not only covers the entire scope of the subject but also explains its philosophy. It will be very useful to students and the subject teachers.

Any suggestion for improvement of the book will be acknowledged, please email your suggestions to manishvarshney@gmail.com.

Manish Varshney
Shanoo Agarwal

Contents

UNIT 2

Chapter 4: Distributed Mutual Exclusion   37

Chapter 5: Distributed Deadlock   50

UNIT 3

Chapter 6: Agreement Protocols   67

UNIT 4

UNIT 5

UNIT 1

1

Introduction to Distributed System

Contents

1.1 Introduction

This certainly is the ideal form of a distributed system, where *implementation detail* of building a powerful system out of many simpler systems is entirely hidden from the user. Are there any such systems? Unfortunately, when we look at the reality of networked computers, we find that the multiplicity of system components usually reflects through the abstractions provided by the operating system and other softwares. In other words, when we work with a collection of independent computers, we are almost always made passively aware of this. For example, some applications require us to identify and distinguish the individual computers by name while in others, the computer hangs due to an error that occurred on a machine that we have never heard of before (Fig. 1.1).

Throughout this course, we will investigate the various technical challenges that ultimately are the cause for the current lack of "true" distributed systems. Moreover, we will investigate various approaches to solve these challenges and study several systems that provide services that are implemented across a collection of computers,

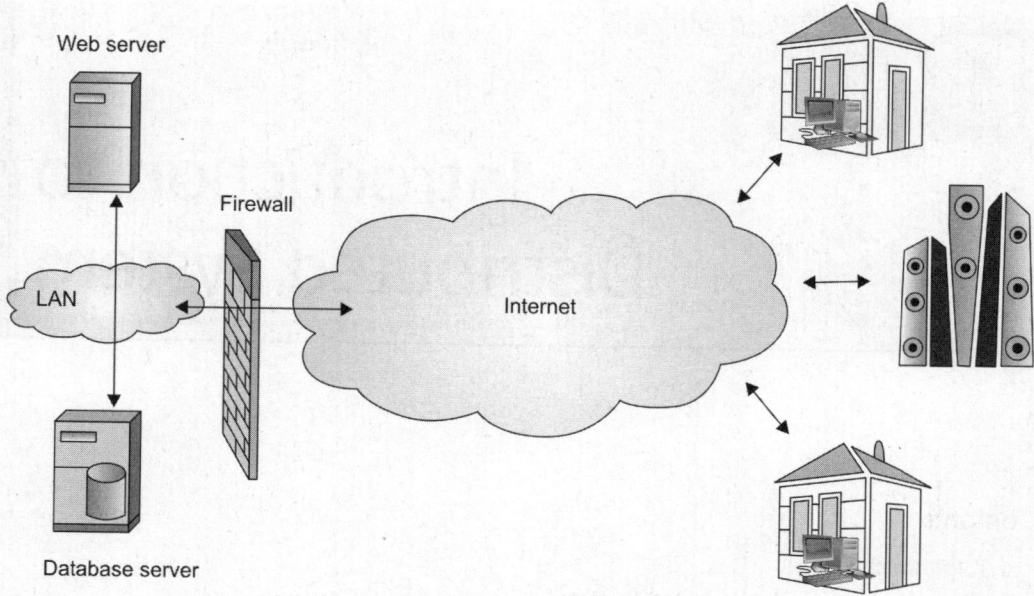

Fig. 1.1: Distributed environment

but appear as a single service to the user. For the purpose of this course, we propose to use the following simple definition of a distributed system:

"A distributed system is a collection of independent computers that are used jointly to perform a single task or to provide a single service."

Distributed system is a collection of independent or separately aparted systems which are connected through network and can communicate only through message passing. They are designed to increase the output and have a fast access. A distributed system consists of multiple software components that are on multiple computers, but run as a single system. The computers in a distributed system can be physically close together and can be connected by a local network or they can be geographically distant and connected by a wide area network (WAN).

1.2 Consequences of Distributed System

1. **Concurrency:** The capacity of the system to handle shared resources can be increased by adding more resources to the network. The two processes are said to be concurrent if either of them can happen before the second one thus interleaving semantics.

2. **No global clock:** The only communication is by sending message through a network, it is not possible to synchronize many computers on network and guarantee synchronization over time, thus the events are logically ordered. However, it is not possible to have a process that can be aware of a single global state.

3. **Independent Failures:** The programs that may not be able to detect whether the network has been failed or has become unusually slow. Running process may be

unaware of other failures within context. Failed processes may go undetected. Both are due to processes running in isolation.

1.3 Definitions of Distributed System

➢ A distributed system (Fig. 1.2) is one in which components located at networked computers communicate and coordinate their actions only through messages.

➢ A distributed system is a collection of independent computers that appears to its users as a single coherent system.

➢ A distributed system has no global clock and also there is no shared memory (Fig. 1.1).

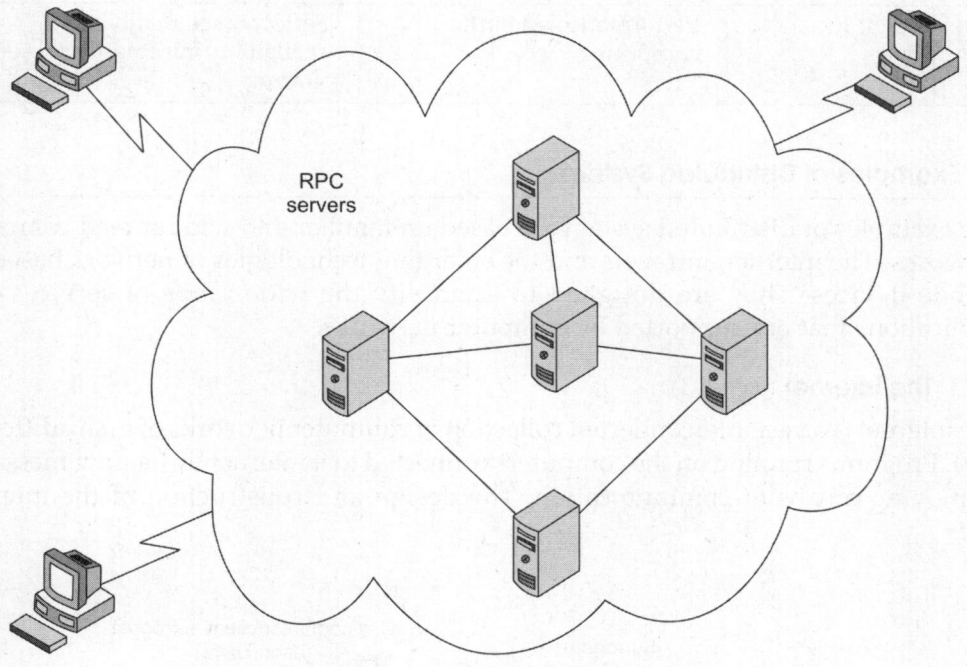

Fig. 1.2: Distributed system

1.4 Distributed System versus Centralized System

S. No	Centralized system	Distributed system
1.	They have non autonomous components	They have autonomous components
2.	They are often manufactured using homogenous technology	They are manufactured using heterogeneous technology
3.	Multiple users share the resources of a centralized system at all times	Distributed system components may be used exclusively and executed in concurrent processes
4.	They have a single point of control and failure	They have multiple points of failure

1.5 Distributed System versus Parallel System

S. No.	Parameters	Parallel system	Distributed system
1.	Memory	Tightly coupled shared memory UMA, NUMA	Distributed memory message passing, RPC and/or use of distributed shared memory
2.	Control	Global clock control SIMD, MIMD	No global clock control, synchronizing algorithms needed
3.	Processor interconnection	Bus, mesh, tree, mesh of tree and hypercube network	Ethernet, token ring and SCI, switching network
4.	Main focus	Performance scientific computing	Performance reliability/availability information/resource sharing

1.6 Examples of Distributed System

The examples of distributed system are based on familiar and wide spread computer networks. The internet, intranets and the emerging technologies of network based on mobile devices. They are designed to exemplify the wide range of services and applications that are supported by computer networks.

1.6.1 The internet (Fig. 1.3)

The internet is a vast interconnected collection of computer networks of many different type. Programs running on the computers connected to it internet by passing messages employing means of communication. The design and construction of the internet

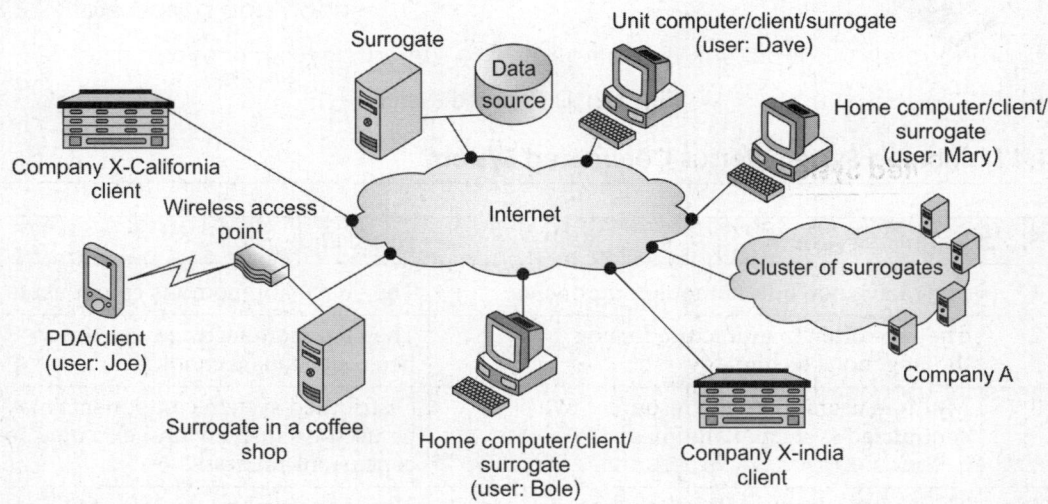

Fig. 1.3: Typical portion of internet

communication mechanism is a major technical achievement, enabling a program running anywhere to address messages to program anywhere else.

1.6.2 The intranet (Fig. 1.4)

An intranet is a portion of the internet that is separately administered and has boundary that can be configured to enforce the security policies. It is composed of several local area networks linked by back bone connection. The network configuration of the particular intranet is the responsibility of the organization that administers it and may vary widely ranging from a LAN on a single site to a connected set of LANs belonging to branches of a company or other organization in different countries. An intranet is connected to internet via router, which allows the users inside the internet to make use of services elsewhere such as web or email. It also allows the users in other intranets to access services it provides. Many organizations need to protect their own services from unauthorizily used by possibly malicious users elsewhere.

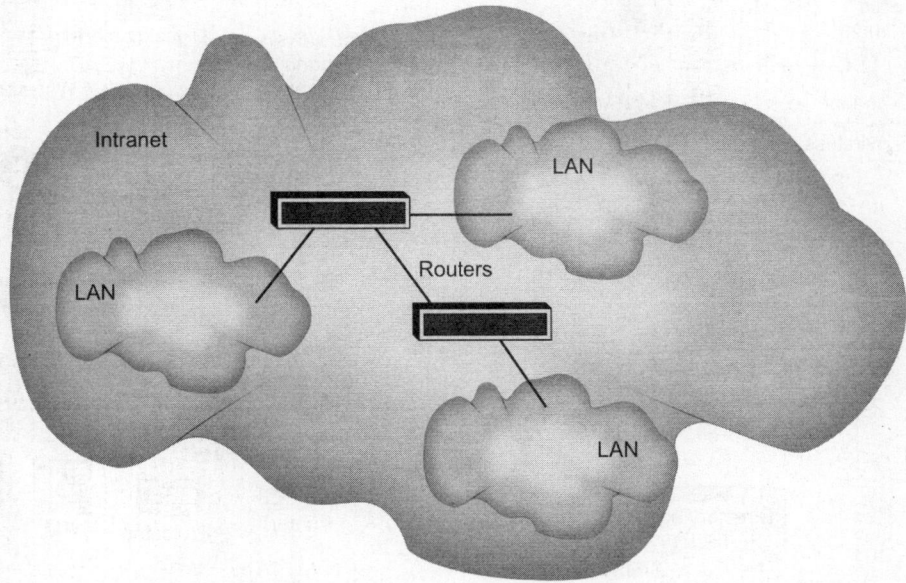

Fig. 1.4: Typical portion of intranet

1.6.3 Mobile and Ubiquitous Computing

Technological advances in device miniaturization and wireless networking have led increasingly to the integration of small and portable computing devices into distributed systems (Fig. 1.5). These devices include:

- Laptop computers
- Handled devices, including personal digital assistants, mobile phones, pagers, video and digital camera.
- Wearable devices, such as smart watches with functionality similar to a PDA.
- Devices embedded in appliances such as washing machines, cars and refrigerators.

The portability of many devices together with their ability to connect conveniently to network in different places makes the mobile computing possible. Mobile computing is the performance of the computing tasks while the users is on the more or visiting places other than their usual environment.

In mobile computing, users who are away from their home intranet are still provided with access to resources via the devices they carry with them. They can continue to access the internet, they can continue to access resources in their home intranet and there is increasing provisions for users to utilize resources such as printers. That are conveniently nearly as they more rounds the latter is also known as location aware computing.

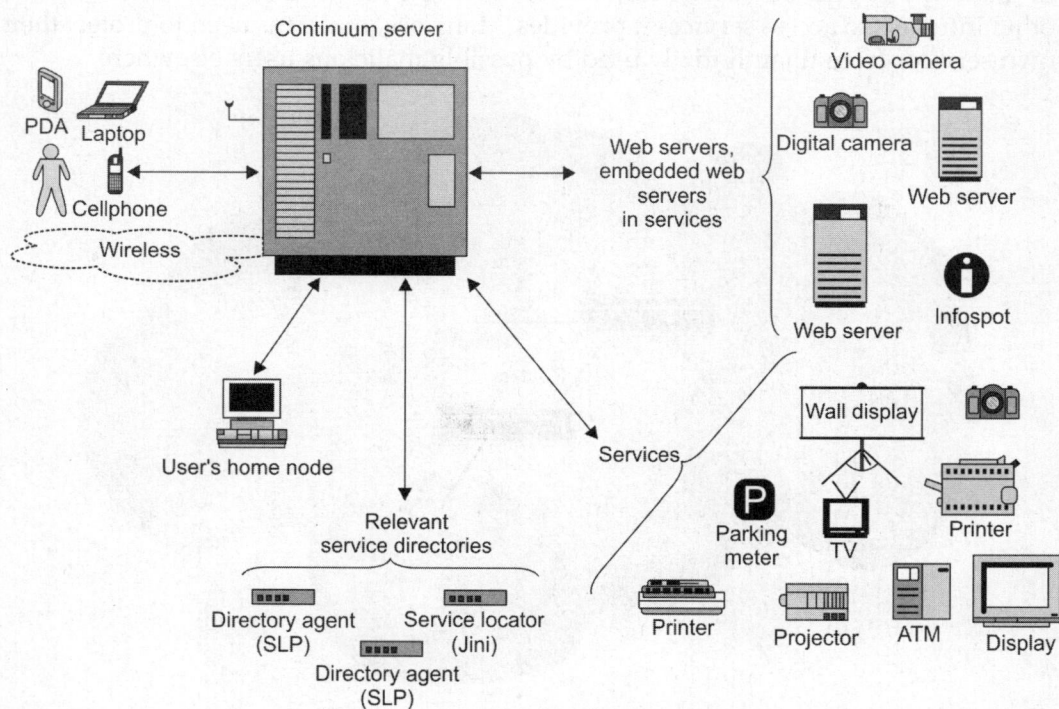

Fig. 1.5: Portable and handheld devices in a distributed system

1.7 Advantages of Distributed System

1. **Performance:** Very often a collection of processors can provide higher performance (and better price/performance ratio) than a centralized computer.

2. **Distribution:** Many applications involve, by their nature, spatially separated machines (banking, commercial, automotive system).

3. **Reliability (fault tolerance):** If some of the machines crash, the system can survive.

4. **Incremental growth:** As requirements on processing power grow, new machines can be added incrementally.

5. **Sharing of data/resources:** Shared data is essential to many applications (banking, computer supported cooperative work, reservation systems); other resources can be also shared (e.g. expensive printers).
6. **Communication:** Facilitates human-to-human communication.

1.8 Disadvantages of Distributed System

1. **Difficulties of developing distributed software:** How should operating systems, programming languages and applications look like?
2. **Networking problems:** Several problems are created by the network infrastructure, which have to be dealt with loss of messages, overloading, and noise over communication path.
3. **Security problems:** Sharing generates the problem of data security.

1.9 Need of Distributed System

- Resource sharing is the main motivation of the distributed system.
- The term "resource" is a rather abstract one, but its best characterizes the range of things that can usefully be shared in a networked computer system.
- Sharing of resources extends from hardware components such as disk and printers of software-defined entities such as files, databases and data objects of all kinds.
- It also includes the stream of video frames and audio connection that a mobile phone represents.

1.10 Design Issues of Distributed System

Design issues that arise *specifically* from the distributed nature of the application:
- Transparency
- Communication
- Performance and scalability
- Heterogeneity
- Openness
- Reliability and fault tolerance
- Security
1. **Transparency**
    - **Access transparency** local and remote resources are accessed using identical operations.
    - **Location transparency** users cannot tell where hardware and software resources (CPUs, files, databases) are located; the name of the resource should not encode the location of the resource.
    - **Migration (mobility) transparency** resources should be free to move from one location to another without having their names changed.
    - **Replication transparency** the system is free to make additional copies of files and other resources (for purpose of performance and/or reliability), without the users noticing.

Example: Several copies of a file; at a certain request that copy is accessed which is the closest to the client.

- *Concurrency transparency* the users will not notice the existence of other users in the system (even if they access the same resources).

- *Failure transparency* applications should be able to complete their task despite failures occurring in certain components of the system.

- *Performance transparency* load variation should not lead to performance degradation. This could be achieved by automatic reconfiguration as response to changes of the load; it is difficult to achieve.

2. **Communication:** Networking infrastructure (interconnections and network software). Appropriate communication primitives and models and their implementation:

- *Communication primitives*
  - send
  - receive
  - remote procedure call (RPC)
- *Communication models*
  - client-server communication: implies a message exchange between two processes: the process which requests a service and the one which provides it;
  - group muticast: the target of a message is a set of processes, which are members of a given group, message passing.

3. **Performance and scalability:** Several factors are influencing the performance of a distributed system:

- The performance of individual workstations.
- The speed of the communication infrastructure.
- Extent to which reliability (fault tolerance) is provided (replication and preservation of coherence imply large overheads).
- Flexibility in workload allocation, for example, idle processors (workstations) could be allocated automatically to a user's task.

Scalability: The system should remain efficient even with a significant increase in the number of users and resources connected:

- cost of adding resources should be reasonable;
- performance loss with increased number of users and resources should be controlled;
- software resources should not run out (number of bits allocated to addresses, number of entries in tables, etc.)

4. **Heterogeneity:** Distributed applications are typically heterogeneous:

- different hardware: Mainframes, workstations, PCs, servers, etc.;
- different software: UNIX, MSWindows, IBM OS/2, Real-time OSs, etc.;

– unconventional devices: Teller machines, telephone switches, robots manufacturing systems.

– diverse networks and protocols: Ethernet, FDDI, ATM, TCP/IP, Novell Netware, etc.

The solution for this is middleware, an additional software layer to mask heterogeneity.

5. **Openness:** One of the important features of distributed systems is openness and flexibility:

– Every service is equally accessible to every client (local or remote);

– It is easy to implement, install and debug new services;

– Users can write and install their own services.

Key aspect of openness:

– Standard interfaces and protocols (like internet communication protocols)

– Support of heterogeneity (by adequate middleware, like CORBA)

6. **Reliability and fault tolerance:** One of the main goals of building distributed systems is improvement of reliability.

Availability: If machines go down, the system should work with the reduced amount of resources.

• There should be a very small number of critical resources; these: resources which have to be up in order the distributed system to work.

• Key pieces of hardware and software (critical resources) should be replicated ⇒ if one of them fails another one takes up—redundancy. Data on the system must not be lost, and copies stored redundantly on different servers must be kept *consistent*.

• The more copies kept, the better the availability, but keeping consistency becomes more difficult.

Fault-tolerance is a main issue related to reliability: the system has to detect faults and act in a reasonable way:

• Mask the fault: Continue to work with possibly reduced performance but without loss of data/information.

• Fail gracefully: React to the fault in a predictable way and possibly stop functionality for a short period, but without loss of data/information.

7. **Security:** Security of information resources:

• *Confidentiality*: Protection against disclosure to unauthorised person

• *Integrity*: Protection against alteration and corruption

• *Availability*: Keep the resource accessible

The appropriate use of resources by different users has to be guaranteed.

1.11 Goals of Distributed System

There are many different types of distributed computing systems and many challenges to overcome in successfully designing one. The main goal of a distributed computing system is to connect users and resources in a transparent, open, and scalable way. Ideally this arrangement is drastically more fault tolerant and more powerful than many combinations of standalone computer systems.

- **Openness**

  Openness is the property of distributed systems such that each subsystem is continually open to interaction with other systems (*see* references). Web Services protocols are standards which enable distributed systems to be extended and scaled. In general, an open system that scales has an advantage over a perfectly closed and selfcontained system. Consequently, open distributed systems are required to meet the following challenges:

- **Monotonicity**

  Once something is published in an open system, it cannot be taken back.

- **Pluralism**

  Different subsystems of an open distributed system include heterogeneous, overlapping and possibly conflicting information. There is no central arbiter of truth in open distributed systems.

- **Unbounded no determinism**

  Asynchronously, different subsystems can come up and go down and communication links can come in and go out between subsystems of an open distributed system. Therefore the time that it will take to complete an operation cannot be bounded in advance.

SUMMARY

Distributed systems are everywhere. The internet enables users throughout the world to access its services wherever they may be located. Each organization manages an intranet, providing local services and internet services for local users and generally provides services to other users in the internet. Small distributed systems can be constructed from mobile computers and small computational devices that are attached to a wireless network.

The consequences of distributed system produce many challenges such as Heterogeneity, openness, security, scalability, failure-handling, concurrency and transparency.

REVIEW QUESTIONS

1. Explain distributed computing?
2. What are the advantages of distributed systems over centralized ones?
3. What are various challenges of distributed system?
4. Explain various examples in distributed systems?
5. What are the various disadvantages of distributed system?

2

System Models

Contents

2.1 Introduction

An architectural model defines the way in which the component of system interact with one another and the way they are mapped onto an underlying network of computer. We described the layered architecture of underlying distributed system software and the architectural model that determine the location and interaction of the component. An architectural model of a distributed system in concerned with the placement of its parts and relationships between them.

2.2 System Model

A distributed system is composed of a number of elements, the most important of which are software components, processing nodes and networks. Some of these elements can be specified as part of a distributed system's design, while others are given (i.e. they have to be accepted as they are). Typically when building a distributed system, the software is under the designer's control. Depending on the scale of the system, the hardware can be specified within the design as well, or already exists and has to be taken in 'as is' manner. The key, however, is that the software components must be distributed over the hardware components in some way.

The software of distributed systems can become fairly complex—especially in large distributed systems—its components can spread over many machines. It is important, therefore, to understand how to organize the system. We distinguish between the logical organisation of software components in such a system and their actual physical organization. The software architecture of distributed systems deals with how software components are organized and how they work together, i.e. communicate with each other. Typical software architectures include the layered, object-oriented, data-centred, and event based architectures. Once the software components are instantiated and placed on real machines, we talk about an actual system architecture. A few such architectures are discussed in this section. These architectures are distinguished from each other by the roles that the communicating processes take on.

Choosing a good architecture for the design of a distributed system allows splitting of the functionality of the system, thus structuring the application and reducing its complexity. Note that there is no single best architecture—the best architecture for a particular system depends on the application's requirements and the environment.

Hence, as per the requirement there are three types of models have been defined, they are:

- *Physical model*: It captures the hardware composition of a system in terms of computers and other devices and their interconnecting network;
- *Architectural model*: It defines the main components of the system, what their roles are and how they interact (software 2 architecture), and how they are deployed in a underlying network of computers (system architecture);
- *Fundamental model*: It gives the formal description of properties that are common to architecture models. Three fundamental models: Interaction models, failure models and security models.

2.3 Architectural Model

An architectural model simplifies and abstracts the functions of the individual components of a distributed system and then it considers (Fig 2.1).
- The placement of the components across a network of computers
- The inter-relationships between the components
- **Process classification**
  - Server process: A process that accepts requests from other processes
  - Client process, peer process

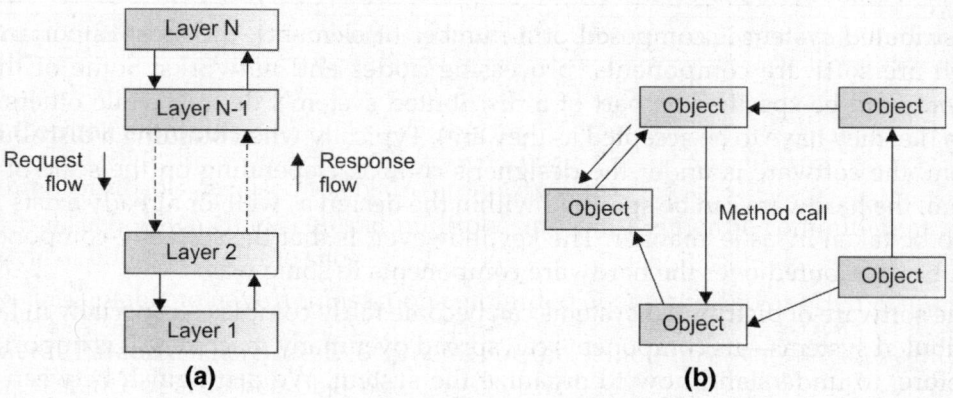

Fig. 2.1: Basic architectural styles

2.3.1 Client–server model (Fig. 2.2)

The system is structured as a set of processes, called servers, that offer services to the users, called clients. The client–server model is usually based on a simple request/reply protocol, implemented with send/receive primitives or using remote procedure calls (RPC) or remote method invocation (RMI):

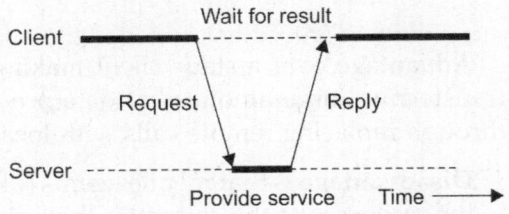

Fig. 2.2: Client–server model

- The client sends a request (invocation) message to the server asking for some service;
- The server does the work and returns a result (e.g. the data requested) or an error code if the work can not be performed.

Variants of client-server model

There are several variations in the client–server model and can be derived from the following factors:

a. The use of mobile code and mobile agent
b. User need for low cost computers with limited hardware resources that are simply to manage.
c. The requirement to add and remove mobile devices in convenient manner.

Mobile code

Applets are an example of mobile code (Fig. 2.3a). In this case, once the downloaded applet runs locally on the client side/web browser it gives better interactive response since network access is subsequently avoided.

Pull versus push model: Most interactions with the web server are initiated by the client to access data (Fig. 2.3b). This is the *pull model*. However for some applications this may not work.

For example, a stock broker's application where the customer needs to be kept informed of any changes in the share prices as they occur at the information source on the server side. In this case we need additional software (may be a special applet) that receives

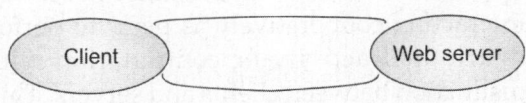

Fig. 2.3a: Applet code

updates from the server. This is the *push model*. The applet would then display the new prices to the user and may be perform automatic buy/sell operations triggered by conditions set up by the customer and stored locally in the customer's computer.

Mobile agents

A mobile agent is a running program (including both code and data) that travels from one computer to another in a network carrying out a task on someone's behalf (such as collecting information), eventually returning with the results. Such an agent may, for example, access the local database.

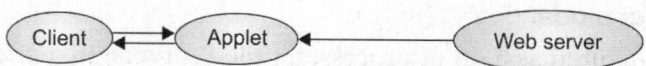

Fig. 2.3b: Client interact with applet

Advantage over a static client making remote method calls on a server, possibly transferring large amounts of data are reduction in communication cost and time through replacing remote calls with local ones.

Disadvantage is that mobile agents (like mobile code) are a potential security threat to the resources of the computer they visit. Need to verify the identity of the user on whose behalf the mobile code is acting (digital signatures) and then provide access (limited or full). The applicability of mobile agents may be limited.

Proxy servers and caches (Fig. 2.4)

Web browsers maintain a cache of recently visited web pages and other web resources in the client's local file system, using a special HTTP request to check with the original server that the cached pages are up to date before displaying them.

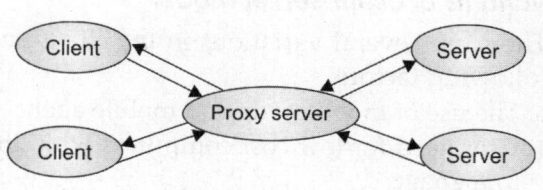

Fig. 2.4: Proxy servers

Web proxy servers provide a shared cache of web resources for the client machines at a site or across several sites. The purpose of the proxy server is to increase availability of the service by reducing the load on the WAN and web servers.

2.3.2 Peer processes (Fig. 2.5)

All processes play similar roles, have similar application and communication code, interacting cooperatively as peers to perform a distributed activity or computation with no distinction between clients and servers. This can reduce IPC delays.

For example, in a whiteboard application that allows several computers to view and interactively modify a picture that is shared between them, each peer process can use middleware to perform event notification and group communication to notify all the other

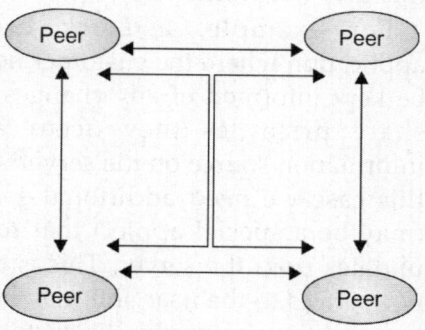

Fig. 2.5: peer to peer model

application processes of changes to the picture. This would provide better interactive response than a server-based architecture where the server would be responsible for broadcasting all updates.

2.4 Fundamental Model

It provides the description of properties that are present in all distributed architectures. It is also an abstraction of essential properties of a natural phenomenon, for the purpose of understanding and analysis.

Purpose of the model is to

- Make explicit all relevant assumptions
- Make decentralizations concerning what is possible or impossible.

The fundamental models are addressed by three different models, they are:

1. *Interaction models*: Issues dealing with the interaction of processes such as performance and timing effect.
2. *Failure models*: Specification of faults that can be exhibited by processes and communication channels.
3. *Security model*: Threats to processes and communication channels.

2.4.1 Interaction Model

Processes in a distributed system (e.g. client-side and server-side processes) interact with each other by passing messages, resulting in communication (message sharing) and coordination (synchronization and ordering of activities) between processes. Each process has its own state. There are two significant factors affecting process interaction in distributed systems:

1. Communication performance is often a limiting characteristic;

2. There is no single global notion of time since clocks on different computers tend to drift.

Performance of communication channels: Communication over a computer network has the following performance characteristics relating to *latency, bandwidth and jitter:*

The delay between the sending of a message by one process and its receipt by another is referred to as *latency*. The latency includes the propagation delay through the media, the frame/message transmission time, and time taken by the operating system communication services (e.g. TCP/IP stack) at both the sending and receiving processes, which varies according to the current load on the operating system. The *bandwidth* of a computer network is the total amount of information that can be transmitted over it in a given time. *Jitter* is the variation in the time taken to deliver a series of messages. This is relevant to real-time and multimedia traffic.

Two variants of the interaction model are the *Synchronous distributed system* and the *Asynchronous distributed system* models.

- *Synchronous distributed systems* are defined to be systems in which the time to execute each step of a process has a known lower and upper bound; each transmitted message is received within a known bounded time; each process has a local clock whose drift rate from real time has a known bound. It is difficult to arrive at realistic values and to provide guarantees of the chosen values.

- *Asynchronous distributed systems* have no bounds on process execution speeds, message transmission delays and clock drift rates. This exactly models the internet, in which there is no intrinsic bound on server or network load and therefore on how long it takes, fro example, to transfer a file using FTP. Actual distributed systems tend to be asynchronous in nature.

2.4.2 Failure model

In a distributed system both processes and communication channels may fail. There are 3 categories of failures: *omission failures, byzantine (or arbitrary) failures,* and *timing failures.*

1. **Omission failures:** These refer to cases when a process or communication channel fails to perform actions that it is supposed to.

   - *Process omission failures:*

     1. *Process crash:* The main omission failure of a process is to crash, i.e. the process has halted and it will not execute any more. Other processes may or may not be able to detect this state. A process crash is detected via timeouts. In an asynchronous system, a timeout can only indicate that a process is not responding, it may have crashed or may be slow, or the message may not have arrived yet.

     2. *Process fail-stop:* A process halts and remains halted. Other processes may detect this state. This can be detected in synchronous systems when timeouts are used to detect when other processes fail to respond and messages are guaranteed to be delivered within a known bounded time.

   - *Communication omission failures:*

     1. *Send-omission failure:* The loss of messages between the sending process and the outgoing message buffer.

     2. *Receive-omission failure:* The loss of messages between the incoming message buffer and the receiving process.

     3. *Channel omission failure:* The loss of messages in between, i.e. between the outgoing buffer and the incoming buffer.

2. **Byzantine or arbitrary failures:** A process continues to run, but responds with a wrong value in response to an invocation. It might also arbitrarily omit to reply. This kind of failure is the hardest to detect. Communication channels can also exhibit this kind of failure by delivering corrupted messages; delivering messages more than once; or deliver non-existent messages. These kind of messages are rare because communication software (e.g. TCP/IP) use checksums to detect corrupted messages and use message sequence numbers to detect non-existent and duplicate messages. Thus this kind of failure is masked either by hiding it or by converting it into a more acceptable type of failure. For example checksums are used to mask corrupted messages—effectively converting a byzantine failure into an omission failure.

3. **Timing failures:** These are applicable only to synchronous distributed systems where time limits are set on process execution time, message delivery time, and clock drift rate. Any of these failures may result in responses being unavailable to

clients within a specified time interval. In asynchronous distributed systems, no timing failures can be said to occur (even if a slow server response causes a timeout) because no timing guarantees have been made.

2.4.3 Security models

The security of a distributed system can be achieved by securing the processes and the channels used for their interactions and by protecting the objects (e.g. web pages, databases, etc.) that they encapsulate against unauthorized access.

Protecting objects: Some objects may hold a user's private data, such as their mailbox, and other objects may hold shared data such as web pages. *Access rights* are used to specify who is allowed to perform which kind of operations (e.g. read / write / execute) on the object.

- Threats to processes (like server or client processes) include not being able to reliably determine the identity of the sender.
- Threats to communication channels include copying, altering, or injecting messages as they traverse the network and its routers. This presents a threat to the privacy and integrity of information. Another form of attack is saving copies of the message and to replay it at a later time, making it possible to reuse the message over and over again (e.g. remove a sum from a bank account).
- Encryption of messages and authentication using digital signatures is used to defeat security threats.

2.5 Resource Sharing and Web

The term resource is a rather abstract one, but it is best characterizes the range of things that can be usefully shared in a networked computer system. It extends from hardware components such as disk, printers to software-defiend entities such as files, databases asnd data objects of all kinds. Resources are shared to reduce cost. Nowadays users are so accustomed to the benefits of resource sharing that they may easily overlook their significant users are concerned with the sharing data in the form of shared database or a set of web pages not the disc or processors (Fig. 2.6).

- Resources may be shared in the form of printer, scanner, machines, etc.

- Sharing of higher level resources.

- The patterns of sharing and to determine what mechanism the system must supply to coordinate user action.

There are various terms which is used in the web are:

a. Service: It is a distinct of a computer system that manages a collection of related resources and present functionality to users.

b. Server: It is probably familiar word to most of users. It means a running program on computer networked, computer that accept request from program running on other computer to perform a service and respond appropriately.

Resources may be encapsulated as objects and accessed by client objects. In this case a client object invokes a method upon a server object.

World Wide Web (WWW)

The WWW is an evolving system for publishing and accessing resources and services across the internet. Web is an open system. Its operations are based on free published communication standards and document standards. The web is one with respect to the types of 'resource' that can be published and shared on.

The web is based on three main standard technological components. They are:

i. *Hypertext markup language (HTML):* It is language for specifying the contents and layout of pages as they are displayed by web browsers.

ii. *Uniform resource locators (URLs):* Which identify documents and other resources stored as part of web.

iii. *Hypertext transfer protocol (HTTP):* A client server system architecture with standard rules for interaction HTTP by which browsers and other clients fetch documents and other resources from web servers.

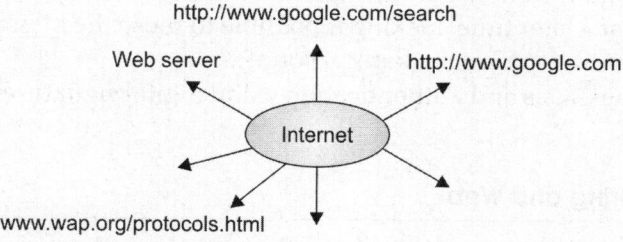

Fig. 2.6: Web servers and browsers

The internet enables users to access services and run applications over a homogeneous collection of computer and networks.

SUMMARY

There are various architectural models given the centralized idea of what purpose the models will be used. For generating applications to send the query to application server and the request may be approved. There are some fundamental models: interaction, failure, security, etc. the threat possibility in each model whether the threat to unauthorized user. There are some mobile codes that run on the web as well as some models provide the quality of service.

Resource sharing is the main motivation for constructing distributed system. Resources such as printers, files, web pages or database records are managed by servers of the appropriate type.

REVIEW QUESTIONS

1. What do you understand by resource sharing?
2. Describe various architectural models.
3. Describe fundamental models.
4. What do you understand by world wide web (WWW)?
5. Write short notes on:
   a. Failure model
   b. Mobile agent
   c. Mobile code
   d. Proxy server

3 Theoretical Foundation for Distributed System

Contents

3.1 Introduction

Distributed system is a collection of separately aparted systems and the communication is done only through message passing and these messages are delivered after a transmission delay. In this chapter types of clocks, limitations of distributed system caused by lack of common memory and a system wide common clock that can be shared by all the processes are described in detail.

3.2 Limitations of Distributed System

There are two inherent limitations of distributed systems which are:
1. Absence of global state
2. Absence of shared memory

3.2.1 Absence of global state

In distributed system there exists no system with common clock, i.e. a global clock. Each system has its own physical clock generally known as logical clock. We can say that a common clock for all computers can solve this problem by process synchronization and is a very easy task. If a distributed system contains a global clock and make it available to all the computers and processes in the system then it might be possible that two different processes observe the same clock in two different instances. This may be caused due to transmission delay. Hence, the process seen on two different instances but in real it is a single instance. So, we can say that it is a useful feature in distributed system.

3.2.2 Absence of shared memory

All computer systems in distributed environment have their own specific memory. So, the entire system does not have any shared memory, because of this limitation the current state of the system is not provided to every user. To solve this problem problem each processor is provided with the system view. This view can be coherent or incoherent when all the processors make their observations at same physical time, we can say this is coherent view or the partial view and incoherent or the complete view is related to local state of all computers and also any message in the transmission state.

3.3 Concepts of Clock

In distributed control systems, the order of occurrences of events that can be established, if a common notion of time is available. Time is established using different clocks.

Clock synchronization

Clock synchronization deals with understanding the temporal ordering of events produced by concurrent processes. It is useful for synchronization between senders and receivers of messages, control of joint activity, and the serialization of concurrent access to shared objects. For these kinds of events, we introduce the concept of a *logical clock*, one where the clock need not have any bearing on the time of day but should be useful for generating message sequence numbers. It is also useful for certain applications to have access to an accurate *physical clock*, one that attempts to provide an accurate measure of the current time.

➤ *Single CPU*: Critical regions, mutual exclusion, and other synchronization problems are solved using methods such as semaphores and monitors.

➤ *Distributed system*: Semaphores and monitors are not appropriate since they rely on the existence of shared memory, problems to be tackled with:
  • Time
  • Mutual exclusion
  • Election algorithms
  • Atomic transactions
  • Deadlocks.

3.4 Physical Clock

Most computers today keep track of the passage of time with a battery-backed up CMOS clock circuit, driven by a quartz oscillator. This allows the timekeeping to take place even if the machine is powered off. When on, an operating system will generally program a timer circuit to generate an interrupt periodically (common times are 60 or 100 times per second). The interrupt service procedure simply adds one to a counter in memory.

The only problem with maintaining a concept of time is when multiple entities attempt to do it concurrently. Two watches hardly ever agree. Computers have the same problem: A quartz crystal on one computer will oscillate at a slightly different frequency than on another computer, causing the clocks to tick at different rates. The phenomenon of clocks ticking at different rates, creating a ever widening gap in perceived time is known as *clock drift*. The difference between two clocks at any point in time is called *clock skew* and is due to both clock drift and the possibility that the clocks may have been set differently on different machines.

3.5 Lamport's Logical Clock

Lamport proposed a scheme the order of event in a distributed system by using logical clock. The execution of the process for the sequence of the event. Due to the absence of synchronized clock and global time in a distributed system, the order in which two events are occurred at two different machines cannot be determined based on the local time at which they occur based on the behavior by the underlying computation.

3.5.1 Happened before relation

The Lamport proposed the logical time between the events of the relative time and we can define happen before relation denoted as follows:

Rule 1: $a \rightarrow b$, if a and b are events in some process and a happened before b.

Rule 2: For any message m,

Send $(m) \rightarrow$ receive (m)

Where send (m) is the event of sending message and receive (m) is event of receiving it.

Rule 3: If a, b and c are the events such that $a \rightarrow b$ and $b \rightarrow c$ then $a \rightarrow c$, i.e. "$\rightarrow$" relation is transitive.

Thus if e and e' are events and if $e \rightarrow e'$ then we can find a series of events $e_1, e_2, e_3, e_4 ... e_n$ occurring at one or more processes either by applying rule 1 or rule 2 of happened before relation. That is either they occur in succession at same process or there is a message m such that e_i = send (m) and e_{i+1} = receive (m).

Causally Related Events

Event a causally affect event b

If $a \rightarrow b$

i.e. they follow 'happened before' relation.

Concurrent Events

Two distinct events a and b are said to be concurrent if $a \rightarrow b$ and $b \rightarrow a$. In other words, they do not causally affect each other, i.e. they do not follow 'happened before' relation. Events can be $a \parallel b$.

3.5.2 Logical clocks

One aspect of clock synchronization is to provide a mechanism whereby systems can assign sequence numbers ("timestamps") to messages upon which all cooperating processes can agree. What matters in many cases is not the time of day but that all processes can agree on the *order* in which related events occur. In this case, our interest is not on obtaining and maintaining true time, but on getting event sequence numbers that make sense system-wide. These clocks are called logical clocks. If processes do not interact then their clocks do not have to be synchronized.

Conditions satisfied by logical clocks

After the happened before relation logical clock follow the following conditions:

Condition 1: For any two event a and b in process P_i, if a occur before b then $(a \rightarrow b)$ $C_i(a) < C_j(b)$.

Condition 2: If a is event message in process P_j then of sending message m in P_i and b is event of receiving $C_i(a) < C_j(b)$.

The following implementation rules (IR) for the clocks guarantee that the clocks satisfy the correctness conditions C_1 and C_2.

[IR-1]: $C_i = C_i + d \ (d > 0)$

[IR-2]: If event a and b are of sending and receiving message m which is assigned a timestamp t_m

i.e. $t_m = C_i(a)$ and follows the following equation to calculate the value

   $C_j = \max (C_j t_m + d)$, $d > 0$; where C_j is calculated from [IR-1].

Example 3.1: Find the Lamport clock value for the following events:

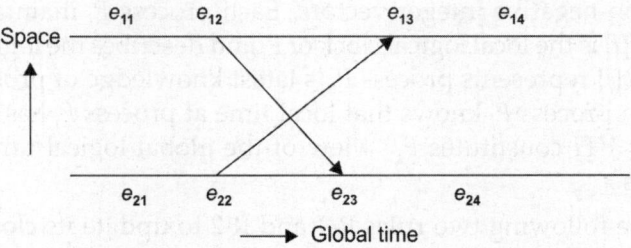

Solution: From rule IR-1 $e_{11} = 1, e_{12} = 1 + 1 = 2$

   $e_{13} = 2 + 1 = 3, e_{14} = 3 + 1 = 4$

   $e_{21} = 1, e_{22} = 1 + 1 = 2$

   $e_{23} = 2 + 1 = 3, e_{24} = 3 + 1 = 4$

From rule IR-2 $C_j = \max (c_j, t_m + d)$

For event e_{13} $C_j = \max (3, 2 + 1) = 3$

For event e_{23} $C_j = \max (3, 2 + 1) = 3$

Limitation of Lamport's Clock

Lamport's system of logical clock, if $a \to b$ then $C(a) < C(b)$. However, the reverse is not necessarily true if the event has occurred in different processes. That if a and b are different processes and $C(a) < C(b)$ then $a \to b$ is not necessarily true; events a and b may be causally related or may not be causally related. Thus, lamport's system of clocks is not powerful enough to capture such situations.

3.5.3 Lamport's algorithm

Each message carries a timestamp of the sender's clock, i.e the sending sequence number according to sender's clock.

- When a message arrives:
  - If receiver's *clock < message timestamp* set system clock to *(message_timestamp + 1)*
  - Else do nothing
- Clock must be advanced between any two events in the same process

The algorithm allows us to proper maintain time ordering among causally-related events. In summary Algorithm needs monotonically increasing software counter. Incremented at least when events that need to be time stamped occur. Each event has a Lamport timestamp attached to it.

For any two events, where $a \to b$:

$$L(a) < L(b)$$

3.6 Vector Clock

The system of vector clocks was developed independently by Fidge, Mattern and Schmuck. In the system of vector clocks, the time domain is represented by a set of n-dimensional non-negative integer vectors. Each process P_i maintains a vector VTI $[1...n]$, where VTI $[i]$ is the local logical clock of P_i and describes the logical time progress at process P_i. VTI $[j]$ represents process P_i 's latest knowledge of process P_j local time. If VTI $[j] = x$, then process P_i knows that local time at process P_j has progressed till x. The entire vector VTI constitutes P_i's view of the global logical time and is used to timestamp events.

Process p_i uses the following two rules R-1 and R-2 to update its clock:

R-1: Before executing an event, process p_i updates its local logical time as follows:
$$VTI \ [i]: = VTI \ [i] + d \ (d > 0)$$

R-2: Each message m is piggybacked with the vector clock vt of the sender process at sending time. On the receipt of such a message (m, vt), process P_i executes the following sequence of actions:

- Update its global logical time as follows:
  $1 \_ k \_ n: VTI \ [k] : = \max \ (VTI \ [k], VT \ [k])$
- Execute R-1.
- Deliver the message m.

The vector also allows us to decide whether two events are causally related or not, i.e. they follow happened before relation or not by simply looking at their timestamp. There are two implementation rules for vector clock they are:

IR-1: Clock c_i is incremented between any two successive events in process

$$P_i\ C_i(i) = c_i(i) + d\ (d > 0)$$

IR-2: If event a is the sending of the message m is assigned a vector timestamp $T_m = c_i(a)$ so c_j will be updated as follow

$$C_j(k) = \max\ (c_j(k),\ t_m(k))$$

On the receipt of message m a process learn about the more recent clock values of the rest of the process in the system.

Example 3.2: Find the vector function of the following

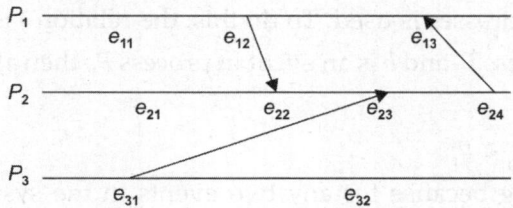

Solution: By the IR-1 for first event of all the process as $C_i(i) = c_i(i) + d$

Initially $c_i\ (i) = 0\ d = 1$

For $c_{11}\ (i) = 0 + 1 = 1$, so vector function of $e_{11} = (1,0,0)$ similarly for $e_{21} = (0,1,0)$ and for $e_{31} = (0,0,1)$

From rule IR-2

For event $e_{12}\ C_i(2) = 1 + 1 = 2$, vector function $= (2,0,0)$

For event $e_{22}\ C_j(k) = \max\ (c_j(k),\ t_m(k))$

$\qquad C_2(1) = \max\ (0,\ 2) = 2$ vector function $= (2,\ 2,\ 0)$

For event $e_{23}\ C_2(3) = \max\ (0,1) = 1$ vector function $= (2,\ 3,\ 1)$

Similarly vector function of $e_{24} = (2,\ 4,\ 1)$

Vector function of $e_{13} = (3,\ 4,\ 1)$

3.7 Total Ordering

We can use the timestamps above and the specified rules to order events by the times that they occur. Ties are broken by process identifier numbers.

The example used is resource contention that must satisfy the following requirements:

- A process must release a resource before the resource can be used by another process.
- Requests must be granted in the same order in which they were made.
- If every process which is granted the resource eventually releases it, then every request is eventually granted.

Clocks following the rules from the previous section can be used to define total ordering of all events. The solution "just involves making sure that each process learns about all other processes' operations.

Several *assumptions* are made:

- Direct connectivity, every process can send messages directly to every other process.
- Messages sent between processes are received in the same order that they were sent.
- Every message is eventually received.

Each process maintains its own request queue which is never seen by any other process.

A system of clocks that satisfy the clock condition can be used to totally order system events. To totally order the events in a system, the events are ordered according to their times of occurrence. In case two or more events occur at the same time, an arbitrary total ordering A of processes is used. To do this, the relation) is defined as follows:

If a is an event in process P_i and b is an event in process P_j, then a) b if and only if either:

i. $C_i(a) < C_j(b)$ or

ii. $C_i(a) = C_j(b)$ and $P_i < P_j$

There is total ordering because for any two events in the system, it is clear which happened first. The total ordering of events is very useful for distributed system implementation.

3.8 Causal Ordering

Causal ordering of messages deals with the concept of maintaining some causal relationship that holds among message send event with corresponding message receive event. It should not be confused with the ordering of events which deals with notion of causal relationship among events. If send $(m_1) \rightarrow$ send (m_2) where send (m_1) is the event of sending message before (m_2) must receive it, then every recipient of both messages M_1 and M_2 must receive M_1 before M_2. In distributed system the causal ordering of messages is not automatically guaranteed. The technique used for causal ordering of messages is useful for the applications such as replicated database so as to maintain consistency of database. It follows two protocols:

1. Birman–Schipher–Stephenson protocol
2. Schiper–Eggli–Sandoz protocol

3.8.1 Birman-Schipher-Stephenson protocol

This protocol uses the principle which allows the handling of message transmitting time, the way to transmit the message, etc. We assume that the compute message is transmitted at once. The process transmitting the message must communicate for broadcasting. There are the following principles applied to this protocol:

a. When the message m is needed to transmit then process P sets the vector time, the vector time of process is denoted by VT_p. It is incremented during message transmission; the timestamp of message is represented by $(VT_p - 1)$.

b. The receiver process P_R receives the message m with its vector timestamp VT_m from the ender process P_S but both the processes are not same, i.e. $P_S \neq P_R$. The process P_R delay its delivery till the following conditions are not satisfied:

i. Vector time for receiver process is equal to the number of messages represented in a unit time, i.e. $VT_p(P_R) = VT_p - 1$

ii. The vector time of process P_R is greater or equal to the vector time of total messages, i.e. $VT_{Pr} \geq VT_m$.

iii. The vector time of receiver process is also updated on receiving a message.

3.8.2 Schiper–Eggli–Sandoz protocol

This protocol is based on process and message transmission. Each process P maintains its own vectors represented by $(V - P)$. Each vector for the process has a vector size of $(N - 1)$, where N is the total number of processes. The processes in the system uses system clock for their ordering. The process ordering is just the placement of process in the queue. There is no requirement of broadcast communication between messages. The protocol uses the logical time for sending the messages also the current logical time. The current logical time is denoted by t_p and logical time at message transfer is denoted by t_m.

3.9 Global State

Recording the global state of a distributed system on the y is an important paradigm. The lack of globally shared memory, global clock and unpredictable message delays in a distributed system make this problem non-trivial.

- The recorded global state may be inconsistent if $n < n'$ or $n > n'$, – where n is the number of messages sent by A along the channel before A's state was recorded. $-n'$ is the number of messages sent by A along the channel before the channel's state was recorded.

    Examples: – Record state of A at state 1, and state of channel and B at state 2, $n = 0$, $n' = 1$

    – Channel at sate 1, state of A and B at state 2, $n' = 0$, $n = 1$.

- Local states – send $(m_{ij}) \in S_{iiff}$ time (send (m_{ij})) < time (LS_i)

    – rec$(m_{ij}) \in S_{jiff}$ time (rec (m_{ij})) < time (LS_j)

    – Transit: transit $(LS_i, LS_j) = \{m_{ij} \mid$ send $(m_{ij}) \in LS_i \wedge$ rec$(m_{ij}) \notin LS_j\}$

    – Inconsistent: inconsistent $(LS_i, LS_j) = \{m_{ij} \mid$ send (m_{ij}) $LS_i \notin$ rec (m_{ij}) "$\in S_j\}$

- Consistent global states: – A global state GS $= \{LS_1, LS_2, ..., LS_n\}$ is consistent iff $\forall_i, \forall_j : 1 \leq i, j \leq n ::$ inconsistent $(LS_i, LS_j) = \Phi$

- Transitless global states: – A global state GS $= \{LS_1, LS_2, ... LS_n\}$ is transitless iff $\forall_i, \forall_j : 1 \leq i, j \leq n ::$ transit $(LS_i, LS_j) = \Phi$

- Strongly consistent global states: It is consistent and transit less

3.9.1 System model

The system consists of a collection of n processes $P_1, P_2, ..., P_n$ that are connected by channels. There are no globally shared memory and physical global clock and processes communicate by passing messages through communication channels. C_{ij} denotes the channel from process P_i to process P_j and its state is denoted by SC_{ij}. The actions performed by a process are modeled as three types of events: Internal events, the message send event and the message receive event. For a message mij that is sent by process P_i to process P_j, let send (m_{ij}) and rec (m_{ij}) denote its send and receive events. At any instant, the state of process p_i, denoted by LS_i, is a result of the sequence of all the events executed by P_i till that instant. For an event e and a process state LS_i, $\in LS_{ii}$ e belongs to the sequence of events that have taken process P_i to state LS_i. For an event e and a process state LS_i, $e \in LS_i$, i.e. does not belong to the sequence of events that have taken process pi to state LS_i. For a channel C_{ij}, the following set of messages can be denied based on the local states of the processes P_i and P_j.

Transit: transit $(LS_i, LS_j) = \{m_{ij} \mid$ send $(m_{ij}) \in LS_i$ V rec(m_{ij}) 6 $\in LS_j$

3.9.2 Models of Communication

Recall, there are three models of communication: FIFO, non-FIFO, and Co. In FIFO model, each channel acts as a rst-in rst-out message queue and thus, message ordering is preserved by a channel. In non-FIFO model, a channel acts like a set in which the sender process adds messages and the receiver process removes messages from it in a random order. A system that supports causal delivery of messages satises the following property: "For any two messages m_{ij} and m_{kj}, if send $(m_{ij})' \rightarrow$ send (m_{kj}), then rec $(m_{ij})' \rightarrow$ rec (m_{kj})".

3.9.3 Consistent global state

The global state of a distributed system is a collection of the local states of the processes and the channels. Notationally, global state GS is denoted as,

$$GS = \{S_i, LS_i, S_i, jSC_{ij}\}$$

A global state GS is a consistent global state i it satisfies the following two conditions:

C-1: send $(m_{ij}) \in LS_i \Rightarrow m_{ij} \in SC_{ij} \oplus$ rec $(m_{ij}) \in LS_j$. ($\oplus$ is Ex-OR operator.)

C-2: send (m_{ij}) 6 $\in LS_i \Rightarrow m_{ij}$ 6 $\in SC_{ij} \wedge$ rec (m_{ij}) 6 $\in LS_j$

3.10 Termination Detection

A fundamental problem: To determine if a distributed computation has terminated. A non-trivial task since no process has complete knowledge of the global state, and global time does not exist. A distributed computation is globally terminated if every process is locally terminated and there is no message in transit between any processes. "Locally terminated" state is a state in which a process has finished its computation and will not restart any action unless it receives a message. In the termination detection problem, a particular process (or all of the processes) must infer when the underlying computation has terminated. A termination detection algorithm is used for this purpose. Messages used in the underlying computation are called basic messages, and messages used for the purpose of termination detection are called control messages. A termination detection (TD) algorithm must ensure the following:

1. Execution of a TD algorithm cannot infinitely delay the underlying computation.

2. The termination detection algorithm must not require addition of new communication channels between processes.

3.10.1 System model

At any given time, a process can be in only one of the two states: active, where it is doing local computation and idle, where the process has (temporarily) finished the execution of its local computation and will be reactivated only on the receipt of a message from another process. An active process can become idle at any time. An idle process can become active only on the receipt of a message from another process. Only active processes can send messages. A message can be received by a process when the process is in either of the two states, i.e. active or idle. On the receipt of a message, an idle process becomes active. The sending of a message and the receipt of a message occur as atomic actions.

3.10.2 Definition of termination detection

Let $P_i(t)$ denote the state (active or idle) of process P_i at instant t.

Let $c_i, j(t)$ denote the number of messages in transit in the channel at instant t from process P_i to process P_j

A distributed computation is said to be terminated at time instant t_0 iff:

$(\forall i:: p_i(t_0) = idle) \wedge (\forall i, j :: c_i, j(t_0) = 0)$.

Thus, a distributed computation has terminated iff all processes have become idle and there is no message in transit in any channel.

3.10.3 Haung's algorithm

System model

A process called controlling agent monitors the computation. A communication channel exists between each of the processes and the controlling agent and also between every pair of processes. Initially, all processes are in the idle state. The weight at each process is zero and the weight at the controlling agent is 1. The computation starts when the controlling agent sends a basic message to one of the processes. A non-zero weight W $(0 < W \leq 1))$ is assigned to each process in the active state and to each message in transit in the following manner:

Basic idea

When a process sends a message, it sends a part of its weight in the message. When a process receives a message, it adds the weight received in the message to its weight. Thus, the sum of weights on all the processes and on all the messages in transit is always 1. When a process becomes passive, it sends its weight to the controlling agent in a control message, which the controlling agent adds to its weight. The controlling agent concludes termination if its weight becomes 1.

Notations

The weight on the controlling agent and a process is in general represented by W.

$B(DW)$—a basic message B sent as a part of the computation, where DW is the weight assigned to it.

$C(DW)$—a control message C sent from a process to the controlling agent, where DW is the weight assigned to it.

Algorithm

The algorithm is defined by the following four rules:

Rule 1: The controlling agent or an active process may send a basic message to one of the processes, say P, by splitting its weight W into W_1 and W_2 such that $W_1 + W_2 = W$, $W_1 > 0$ and $W_2 > 0$. It then assigns its weight $W: = W_1$ and sends a basic message B $(DW: = W_2)$ to P.

Rule 2: On the receipt of the message B (DW), process P adds DW to its weight W $(W: = W + DW)$. If the receiving process is in the idle state, it becomes active.

Rule 3: A process switches from the active state to the idle state at any time by sending a control message $C(DW:=W)$ to the controlling agent and making its weight $W:= 0$.

Rule 4: On the receipt of a message C (DW), the controlling agent adds DW to its weight $(W: = W + DW)$. If $W = 1$, then it concludes that the computation has terminated.

Correctness of algorithm

Notations:

A: Set of weights on all active processes.
B: Set of weights on all basic messages in transit.
C: Set of weights on all control messages in transit.
W_c: Weight on the controlling agent.

Two invariants I-1 and I-2 are defined for the algorithm:

I-1 : $W_c + XW \in (A \cup B \cup C)$
$W = 1$
I-1 : $W_c + XW \in (A \cup B \cup C)$

Invariant I-1 states that sum of weights at the controlling process, at all active processes, on all basic messages in transit, and on all control messages in transit is always equal to 1.

Invariant I-2 states that weight at each active process, on each basic message in transit, and on each control message in transit is non-zero.

Hence, $\qquad WC = 1 \Rightarrow P$

$W \in (A \cup B \cup C) W = 0 \Rightarrow$ (by I-1) $= (A \cup B \cup C) = \varphi$ (by I-2) $(A \cup B) = \varphi$

$(A \cup B) = \varphi$ implies the computation has terminated. Therefore, the algorithm never detects a false termination. Further,

$$(A \cup B) = \varphi \, W_c + P$$

$W \in CW = 1$ (by I-1)

Since the message delay is finite, after the computation has terminated, eventually $W_c = 1$.

Thus, the algorithm detects a termination in finite time.

SUMMARY

Two basic characteristics of a distributed system—the absence of global time and the absence of shared memory—were the main focus.

Two schemes namely Lamport's logical clocks and vector clocks to order events in a distributed system and their usefulness in designing distributed algorithms.

Though a simple bank account example. It has shown how difficult it is to reason about the global state of a system in the absence of shared memory and perfectly synchronized clocks.

We then described the concept of global state and also to collect the consistent set of global state. We also discussed about termination detection and the algorithm required dealing with the termination detection problems.

REVIEW QUESTIONS

1. Discuss about the limitation of distributed systems.

2. Explain virtual time and vector clock.

3. Explain Lamport's logical clock and its limitation.

4. What is the need of ordering of messages in distributed systems?

5. Briefly explain about causal events in the distributed system.

6. Explain the concept of global state and consistent set of global state.

7. Write short notes on:

   a. Active states

   b. Collective global state

   c. Termination detection

8. Explain the algorithm required for termination detection.

9. What is the need of timestamps in distributed system?

UNIT 2

4

Distributed Mutual Exclusion

Contents

4.1 Introduction

Mutual exclusion, concurrent access to a shared resource by several uncoordinated user-requests is arranged to secure the integrity of the shared resource. It requires the action performed by a user on a shared resource must be atomic.

The problem of mutual exclusion frequency arises in distributed system, whenever concurrent access to share resource for several sites is involved. For correctness, if it is necessary that the shared resources can be accessed by a single site or process at a time.

Mutual exclusion ensures that concurrent processes make a serialized access to shared resources or data. The well known critical section problem, in a distributed system neither shared variables (semaphores) nor a local kernel can be used in order to implement mutual exclusion. Thus, mutual exclusion has to be based exclusively on message passing, in the context of unpredictable message delays and no complete knowledge of the state of the system. Sometimes the resource is managed by a server which implements its own lock together with the mechanisms needed to synchronize access to the resource, mutual exclusion and the related synchronization are transparent for the process accessing the resource. This is typically the case for database systems with transaction processing. Often there is no synchronization built in which implicitly protects the resource (files, display windows, peripheral devices, etc.), a mechanism has to be implemented at the level of the process requesting for access (Fig. 4.1).

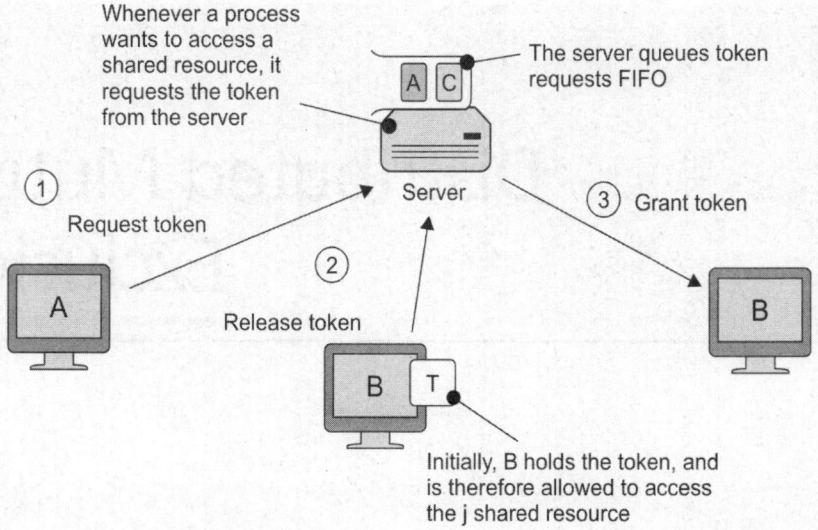

Fig. 4.1: Mutual exclusion in distributed system

Basic requirements for a mutual exclusion mechanism:

- **Safety**: At most one process may execute in the critical section (CS) at a time;
- **Liveness**: A process requesting entry to the CS is eventually granted it (so long as any process executing the CS eventually leaves it). Liveness implies freedom of deadlock and starvation.

System Model

- The system consists of N sites, $S_1, S_2, ..., S_N$.
- We assume that a single process is running on each site. The process at site S_1 is denoted by P_i.
- A site can be in one of the following three states: Requesting the CS, executing the CS, or neither requesting nor executing the CS (i.e. idle).
- In the 'requesting the CS' state, the site is blocked and cannot make further requests for the CS. In the 'idle' state, the site is executing outside the CS.
- In token-based algorithms, a site can also be in a state where a site holding the token is executing outside the CS (called the idle token state).
- At any instant, a site may have several pending requests for CS. A site queues up these requests and serves them one at a time.

4.2 Mutual Exclusion in Single System versus Distributed System

In single computer system the status of shared resources and the status of user is readily available in the shared memory solution to the mutual exclusion problem can be easily implemented using shared variables.

But in distributed system both the shared resource and the users may be distributed and shared memory does not exist. So, approach based on shared variable is not applicable to distributed system. An approach based on message passing must be used. The problem of mutual exclusion become much complex in distributed system, because of the lack of shared memory and a common physical clock and also due to unpredictable message delay. It is virtually impossible for a site in distributed system to have current and complete knowledge of state of system.

4.3 Classification of Mutual Exclusion Algorithms

The problem of mutual exclusion has received considerable attention and several distributed algorithms to achieve mutual exclusion in distributed systems have been proposed. They tend to differ in their communication topology in the amount of information maintained by each site about other sites. These algorithms can be grouped broadly into two classes which are:

1. **Nontoken-based:** Each process freely and equally competes for the right to use the shared resource; requests are arbitrated by a central control site or by distributed agreement.

2. **Token-based:** A logical token representing the access right to the shared resource is passed in a regulated fashion among the processes; whoever holds the token is allowed to enter the critical section.

4.4 Requirements of Mutual Exclusion Algorithms

Primary objective of all algorithms is to guarantee that only one request will access the critical section at a time. The characteristics which should be kept in mind are:

• **Freedom from deadlocks:** Two or more sites should not endlessly wait for messages that will never arrive.

• **Freedom from starvation:** A site should not be forced to wait indefinitely to execute CS while other sites are repeatedly executing CS. That is, every requesting site should get an opportunity to execute CS in finite time.

• **Fairness:** Each process gets a fair chance to execute the CS. Fairness property generally means the CS execution requests are executed in the order of their arrival (time is determined by a logical clock) in the system.

• **Fault tolerance:** A mutual exclusion algorithm is fault tolerant if in the wake of a failure it can reorganize itself so that it continues to function without any disruptions.

4.5 Nontoken-based Algorithms

• Two or more successive rounds of messages are exchanged among the sites to determine which site will enter the CS next.
• Each process freely and equally competes for the right to use the shard resource; requests are arbitrated by a central control site or by distributed agreement.
• Permission from all processes

4.5.1 Lamport's algorithm

➢ **Requirements:**

– A process in the CS must release it before it can be granted to another.

– Requests to enter CS should be granted in the order in which they were made.

– If each use of CS is finite, each request will eventually be met.

➢ **Network assumptions:**

– FIFO channel, and channel is reliable

– Complete connection

➢ N sites S_1, S_2,.,.S_n, each keeps a queue (*request_queue(i)*) of requests for entering the CS

The Algorithm

- Requesting the critical section (Fig. 4.2a):

  1. When a site S_i wants to enter the CS, it broadcasts a request (t_{si}, i) message to all other sites and places the request on request queue$_i$. $((t_{si}, i)$ denotes the timestamp of the request.)

  2. When a site S_j receives the request (t_{si}, i) message from site S_i, places site S_i's request on request queue j and it returns a timestamped REPLY message to S_i.

- Executing the critical section: Site S_i enters the CS (Fig. 4.2b) when the following two conditions hold:

  1. L-1: S_i has received a message with timestamp larger than (ts_i, i) from all other sites.

  2. L-2: S_i 's request is at the top of request queue$_i$.

- Releasing the critical section (Fig. 4.2c):

  1. Site S_i, upon exiting the CS, removes its request from the top of its request queue and broadcasts a timestamped RELEASE message to all other sites.

  2. When a site S_j receives a RELEASE message from site S_i, it removes S_i's request from its request queue.

When a site removes a request from its request queue, its own request may come at the top of the queue, enabling it to enter the CS.

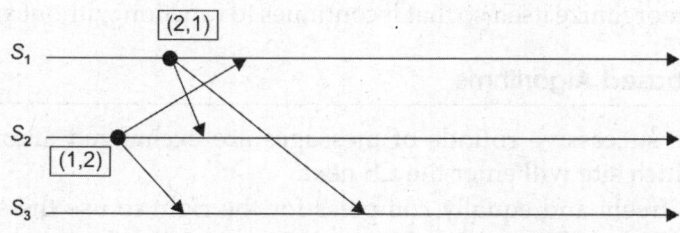

Fig. 4.2a: Making request to enter CS

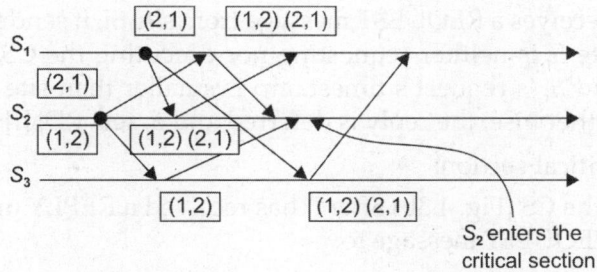

Fig. 4.2b: Entering the CS

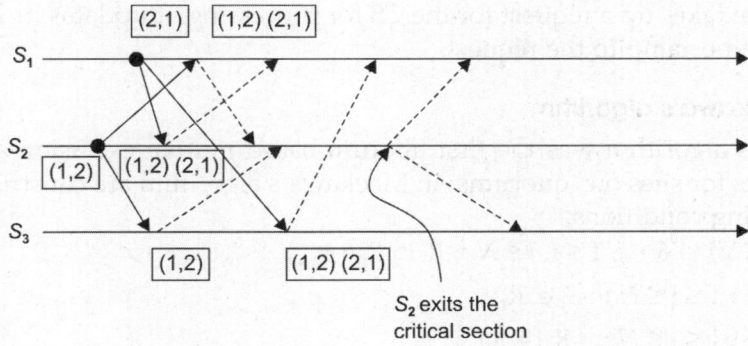

Fig. 4.2c: Releasing the CS

4.5.2 Ricart–Agrawala's algorithm

➤ The Ricart–Agrawala's algorithm assumes the communication channels are FIFO. The algorithm uses two types of messages: REQUEST and REPLY.

➤ A process sends a REQUEST message to all other processes to request their permission to enter the critical section. A process sends a REPLY message to a process to give its permission to that process.

➤ Processes use Lamport-style logical clocks to assign a timestamp to critical section requests and timestamps are used to decide the priority of requests.

➤ Each process P_i maintains the request: Deferred array, RD_i, the size of which is the same as the number of processes in the system.

➤ Initially, $\forall i \, \forall j : RD_i[j] = 0$. Whenever P_i defer the request sent by P_j, it sets $RD_i[j] = 1$ and after it has sent a REPLY message to P_j, it sets $RD_i[j] = 0$.

The algorithm

• Requesting the critical section:

1. When a site S_i wants to enter the CS (Fig. 4.3a), it broadcasts a timestamped REQUEST message to all other sites.

2. When site S_j receives a REQUEST message from site Si, it sends a REPLY message to site S_i if site S_j is neither requesting nor executing the CS, or if the site S_j is requesting and S_i 's request's timestamp is smaller than site S_j's own request's timestamp. Otherwise, the reply is deferred and S_j sets RDj [i]=1

- Executing the critical section:
  - Site S_i enters the CS (Fig. 4.3b) after it has received a REPLY message from every site it sent a REQUEST message to.
- Releasing the critical section (Fig. 4.3c):
  - When site S_i exits the CS, it sends all the deferred REPLY messages: $\forall$j if $RD_i [j]=1$, then send a REPLY message to S_j and set $RD_i [j]=0$.

When a site receives a message, it updates its clock using the timestamp in the message. When a site takes up a request for the CS for processing, it updates its local clock and assigns a timestamp to the request.

4.4.3 Maekawa's algorithm

Maekawa's algorithm was the first quorum-based mutual exclusion algorithm. The request sets for sites (i.e. quorums) in Maekawa's algorithm are constructed to satisfy the following conditions:

M-1: $(\forall i \ \forall j : i \ 6 = j, 1 \le i, j \le N :: R_i \cap R_j \ 6 = \_)$

M-2: $(\forall i : 1 \le i \le N :: S_i \in R_i)$

M-3: $(\forall i : 1 \le i \le N :: |R_i| = K)$

M4: Any site S_j is contained in K number of R_i S, $1 \le i, j \le N$.

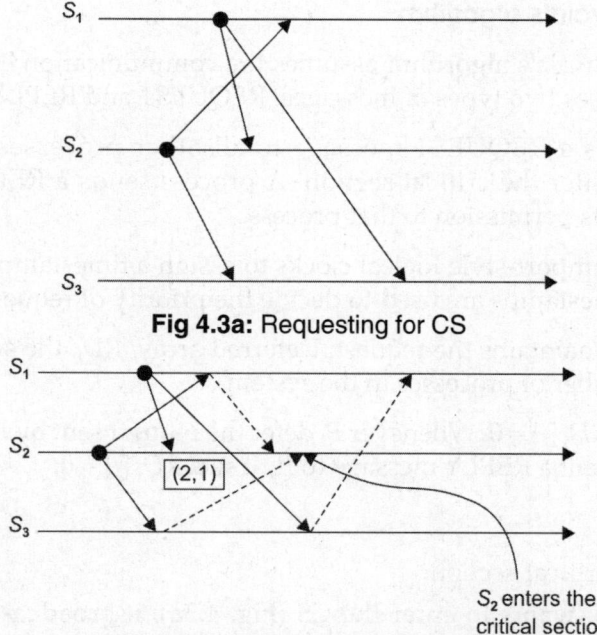

Fig 4.3a: Requesting for CS

S_2 enters the critical section

Fig. 4.3b: Entering the critical section

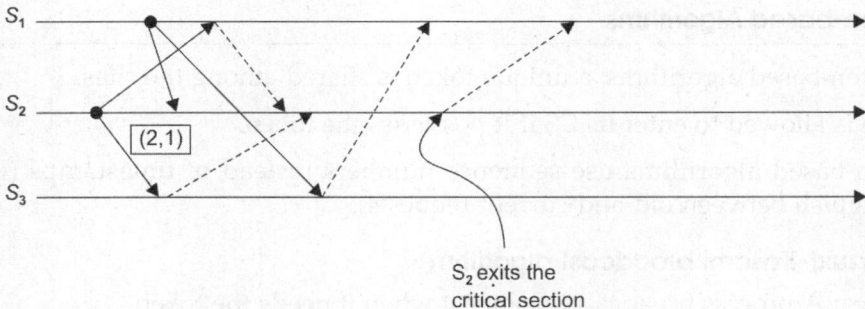

Fig. 4.3c: Releasing the critical section

Maekawa used the theory of projective planes and showed that $N = K(K - 1) + 1$. This relation gives $|R_i| = \sqrt{N}$.

➢ Conditions M-1 and M-2 are necessary for correctness; whereas conditions M-3 and M-4 provide other desirable features to the algorithm.

➢ Condition M-3 states that the size of the requests sets of all sites must be equal implying that all sites should have to do equal amount of work to invoke mutual exclusion.

➢ Condition M-4 enforces that exactly the same number of sites should request permission from any site implying that all sites have "equal responsibility" in granting permission to other sites.

The algorithm

- Requesting the critical section

  1. A site S_i requests access to the CS by sending REQUEST (i) messages to all sites in its request set R_i.

  2. When a site S_j receives the REQUEST (i) message, it sends a REPLY(j) message to Si provided it hasn't sent a REPLY message to a site since its receipt of the last RELEASE message. Otherwise, it queues up the REQUEST(i) for later consideration.

- Executing the critical section

  1. Site S_i executes the CS only after it has received a REPLY message from every site in R_i.

- Releasing the critical section

  1. After the execution of the CS is over, site S_i sends a RELEASE (i) message to every site in R_i.

  2. When a site S_j receives a RELEASE (*i*) message from site S_i, it sends a REPLY message to the next site waiting in the queue and deletes that entry from the queue. If the queue is empty, then the site updates its state to reflect that it has not sent out any REPLY message since the receipt of the last RELEASE message.

4.6 Token-based Algorithms

- In token-based algorithms, a unique token is shared among the sites.
- A site is allowed to enter its CS if it possesses the token.
- Token-based algorithms use sequence numbers instead of timestamps (Used to distinguish between old and current requests).

4.6.1 Suzuki–Kasami broadcast algorithm

Basic idea: A process broadcasts a request when it needs the token.

- The process which has the token sends it to the requesting process after it finishes the CS.
- The Token
  - Q: Queue of the requesting processes, at most n.
  - LN $[1,...,n]$: array of integers, $LN[j]$ is the sequence number of the request that S_j executed most recently.

The algorithm

- Requesting the critical sections
  1. If requesting site S_i does not have the token, then it increments its sequence number, RN_i $[i]$, and sends a REQUEST(i, sn) message to all other sites. ('sn' is the updated value of RN_i $[i]$.)
  2. When a site S_j receives this message, it sets RN_j $[i]$ to max (RN_j $[i]$, sn). If S_j has the idle token, then it sends the token to S_i if RN_j $[i]$ = $LN[i]$ + 1.
- Executing the critical section:
  1. Site S_i executes the CS after it has received the token.
- Releasing the critical section: Having finished the execution of the CS, site S_i takes the following actions:
  1. It sets $LN[i]$ element of the token array equal to RN_i $[i]$.
  2. For every site S_j whose id is not in the token queue, it appends its id to the token queue if RN_i $[j]$ = $LN[j]$ + 1.
  3. If the token queue is nonempty after the above update, S_i deletes the top site id from the token queue and sends the token to the site indicated by the id.

4.6.2 Singhal's heuristic algorithm

In this algorithm, each site maintains about the state of other sites in the system and uses it to select a set of sites that are likely to have token. The site requests the token only from the selected sites. Hence, reducing the number of messages to execute critical section. It is called heuristic algorithm because the sites are selected for sending request message heuristically.

S_i maintains SV_i $[1,..., N]$ and SN_i $[1,..., N]$ for storing information on other sites: state and highest sequence number.

- Token contains 2 arrays: TSV $[1,..., N]$ and TSN $[1,..., N]$.
- States of a site
  - R: Requesting CS

- E: Executing CS
- H: Holding token, idle
- N: None of the above

- Initialization:
  - $SV_i[j] := N$, for $j = N,...,$ i; $SV_i[j] := R$, for $j = i - 1, 1$; $SN_i[j] := 0, j = 1,...,N$. S-1 (site 1) is in state H.
  - Token: $TSV[j] := N$ and $TSN[j] := 0, j = 1,...,N$.

The algorithm

- Requesting the critical section

  1. If S_i has no token and requests CS:
     - $SV_i[i] := R$. $SN_i[i] := SN_i[i] + 1$.
     - Send REQUEST(i, sn) to sites S_j for which $SV_i[j] = R$. (sn: sequence number, updated value of $SN_i[i]$).

  2. Receiving REQUEST(i, sn): if $sn \Leftarrow SN_j[i]$, ignore. Otherwise, update $SN_j[i]$ and do:
     - $SV_j[j] = N \rightarrow SV_j[i] := R$.
     - $SV_j[j] = R \rightarrow$ If $SV_j[i] \mathrel{!=} R$, set it to R and send REQUEST($j, SN_j[j]$) to S_i. Else do nothing.
     - $SV_j[j] = E \rightarrow SV_j[i] := R$.
     - $SV_j[j] = H \rightarrow SV_j[i] := R$, $TSV[i] := R$, $TSN[i] := sn$, $SV_j[j] = N$. Send token to S_i.

- Executing the critical section: After getting token. Set $SV_i[i] := E$.

- Releasing CS

  1. $SV_i[i] := N$, $TSV[i] := N$. Then, do:
     - For other S_j: if $(SN_i[j] > TSN[j])$, then $\{TSV[j] := SV_i[j]; TSN[j] := SN_i[j]\}$
     - else $\{SV_i[j] := TSV[j]; SN_i[j] := TSN[j]\}$

  2. If $SV_i[j] = N$, for all j, then set $SV_i[i] := H$. Else send token to a site S_j provided $SV_i[j] = R$.

4.6.3 Raymond's tree algorithm

➤ Sites are arranged in a logical directed tree. Root: Token holder. Edges: Directed towards root (Fig. 4.4).

➤ Every site has a variable *holder* that points to an immediate neighbour node, on the directed path towards root (root's holder point to itself).

- Requesting the critical section (Fig. 4.5a):

  1. To enter the CS, a site sends a request to the node along the path to the root if it does not hold the token and its request_q is empty. It adds its request to its request_q.

2. When a site on the path receives the message, it places the request in its request_q and sends a request along the path to the root if it has not sent out a request.

3. When the root receives a request, it sends the token to the site from which it received the request and sets its holder to point at that site.

4. When a site receives the token, it deletes the top entry from its request_q, sends the token to the site indicated in this entry, and sets its holder to point at that site. If request_q is nonempty, it sends a request to the site which is pointed at by holder (Fig. 4.5b).

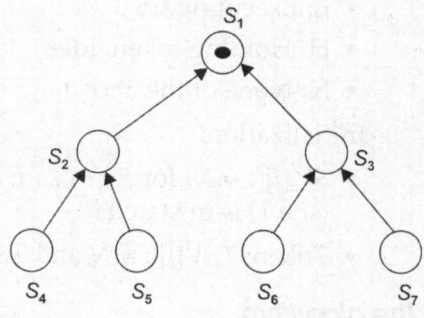

Fig. 4.4: Tree

- Executing the critical section:
  - A site enters the CS when it receives the token and its own request is at the top of its request_q. In this case, remove the top entry.
- Releasing the critical section:
  1. If its request_q is nonempty, it deletes the top entry from its request_q, sends the token to that site, and sets its holder to that site.
  2. If the request_q is nonempty at this point, it sends a request to the site which is pointed at by the holder.

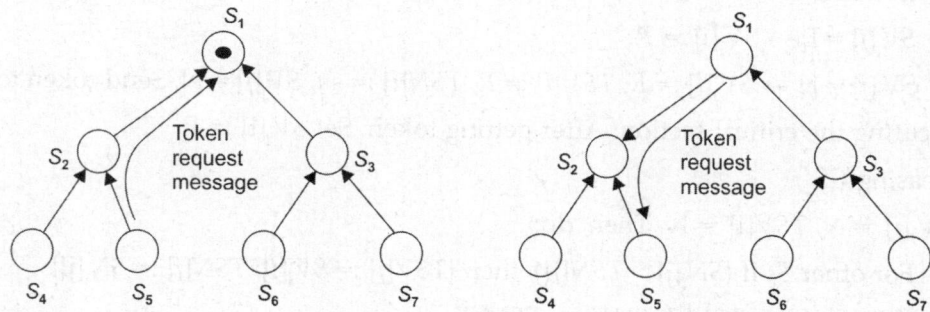

Fig. 4.5a: Requesting the critical section

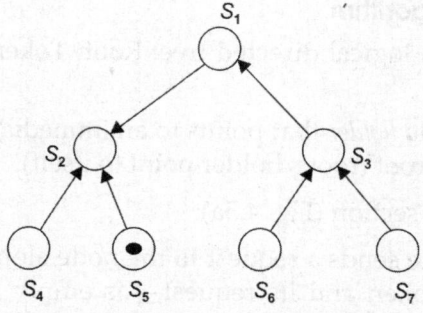

Fig. 4.5b: Receiving the token

4.7 Performance Metrics

The performance is generally measured by the following four metrics:

1. Message complexity: The number of messages required per CS execution by a site.

2. Synchronization delay (Fig. 4.6a): After a site leaves the CS, it is the time required and before the next site enters the CS.

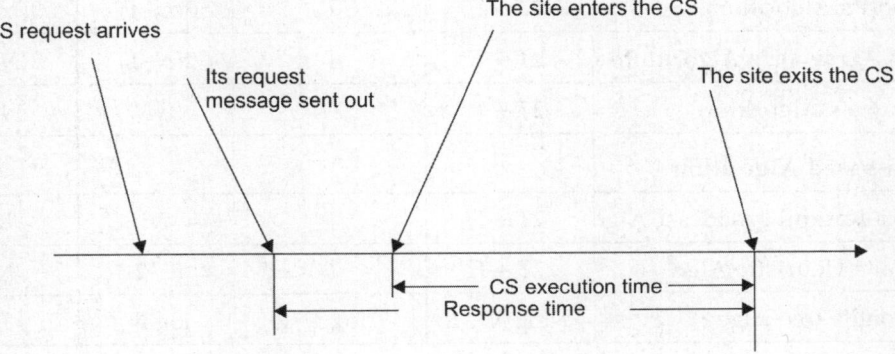

Fig. 4.6a: Synchronization delay

3. Response time: The time interval a request waits for its CS execution to be over after its request messages have been sent out.

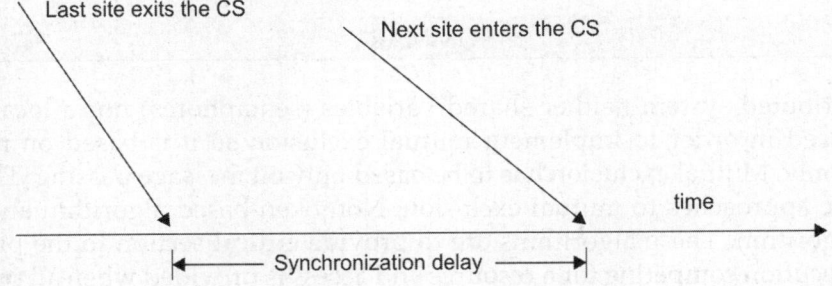

Fig. 4.6b: Response time

4. System throughput: The rate at which the system executes requests for the CS. System throughput = 1/(SD + E), where SD is the synchronization delay and E is the average critical section execution time.

- **Low and high load performance:**
  1. We often study the performance of mutual exclusion algorithms under two special loading conditions, viz. "low load" and "high load".
  2. The load is determined by the arrival rate of CS execution requests.
  3. Under low load conditions, there is seldom more than one request for the critical section present in the system simultaneously.
  4. Under heavy load conditions, there is always a pending request for critical section at a site.

4.6.1 Performance matrix for mutual exclusion algorithm

	Response time	Synchronization delay	Message (low load)	Message (high load)
Nontoken-based Algo				
Lamport's Algorithm	$2T + E$	T	$3(N-1)$	$3(N-1)$
Ricart Agrawala's Algorithm	$2T + E$	T	$2(N-1)$	$2(N-1)$
Maekawa's Algorithm	$2T + E$	$2T$	$3\sqrt{N}$	$5\sqrt{N}$
Token-based Algorithm				
Suzuki–Kasami broadcast Algo	$2T + E$	T	N	N
Singhal's Heuristic Algo	$2T + E$	T	$N/2$	N
Raymond's tree Algo	$T \log N + E$	$T \log N/2$	$\log N$	T

Where,

$N \rightarrow$ No. of processes

$T \rightarrow$ Message transmission time

$E \rightarrow$ Critical section execution time

SUMMARY

In a distributed system neither shared variables (semaphores) nor a local kernel can be used in order to implement mutual exclusion so it is based on message passing only. Mutual exclusion has to be based only on message passing. There are two basic approaches to mutual exclusion: Nontoken-based algorithm and token based algorithm. These algorithms are to provide critical section to the processes for its execution competing for a resource and access is provided when all processes have replies to the request. The algorithms are expensive in terms of message traffic and failure of any process prevents progress. For many distributed applications it is needed that one process acts as a coordinator, an election algorithm has to choose one and only one process from a group, to become the coordinator. All group members have to agree on the discussion. Processes are allowed to fail during election process.

REVIEW QUESTIONS

1. What do you understand by distributed mutual exclusion?
2. Explain various nontoken-based algorithm
   i. Ricart–Agrawala's algorithm

     ii. Lamport's algorithm

3. How do you classify distributed mutual exclusion?

4. Explain various non-token based algorithm

     i. Suzuki–Kasami broadcast algorithm

     ii. Raymond's tree algorithm

5. What do you understand by distributed mutual exclusion? Many distributed algorithm require the use of a coordinating process. To what extent can such algorithm actually be considered distributed are discussed?

5

Distributed Deadlock

Contents

5.1 Introduction

Deadlock can occur whenever two or more processes are competing for limited resources and the processes are allowed to acquire and hold a resource (obtain a lock), thus preventing others from using the resource while the process waits for other resources. Two common places where deadlocks may occur are with processes in an operating system (distributed or centralized) and with transactions in a database. The concepts discussed here are applicable to any system that allocates resources to processes. Locking protocols such as the popular *two phase locking* (see concurrency control) give rise to deadlock as follows: process A gets a lock on data item X while process B gets a lock on data item Y. Process A then tries to get a lock on Y. As Y is already locked, process A enters a blocked state. Process B now decides to get a lock on X, but is blocked. Both processes are now blocked, and by the rules of two phase locking, neither will relinquish their locks.

Process A is waiting for a resource held by process B and process B is waiting for a resource held by process A. No progress will take place without outside intervention. Several processes can be involved in a deadlock when there exists a cycle of processes waiting for each other. Process A waits for B which waits for C which waits for A.

Four conditions must hold for deadlock to occur:

1. *Exclusive use*— when a process accesses a resource, it is granted an exclusive use of that resource.

50

2. *Hold and wait*—a process is allowed to hold on to some resources while it is waiting for other resources.
3. *No pre-emption*—a process cannot preempt or take away the resources held by another process.
4. Cyclical wait—there is a circular chain of waiting processes, each waiting for a resource held by the next process in the chain.

5.1.1 Distributed deadlock

Deadlocks in distributed systems are similar to deadlocks in centralized systems. In centralized systems, we have one operating system that can oversee resource allocation and aware whether deadlocks are (or will be) present. With distributed processes and resources, it becomes harder to detect, avoid, and prevent deadlocks. Several strategies can be used to handle deadlocks:

Ignore: we can ignore the problem. This is one of the most popular solutions.

Detect: we can allow deadlocks to occur, then detect that we have a deadlock in the system, and then deal with the deadlock.

Prevent: we can place constraints on resource allocation to make deadlocks impossible.

Avoid: we can choose resource allocation carefully and make deadlocks impossible. Deadlock avoidance is never used (either in distributed or centralized systems). The problem with deadlock avoidance is that the algorithm will need to know resource usage requirements in advance so as to schedule them properly.

The structure of a system may allow additional complexities in the deadlock problem. The simplest model, single resource, requires that a process have no more than one unfulfilled request. Thus, a blocked process is waiting for only one other process and can be involved in at most one deadlock cycle.

In the AND model (also called the multiple resource model), a process is allowed to make several resource requests, and it is blocked until all of the requests are granted. Processes in this model can be involved in several deadlock cycles at once. In the OR model (also called the communication model), a process makes several requests and is blocked until any one of them is granted. The AND–OR model allows a combination of request types, such as a request for resource X and either Y or Z. Unless specifically stated, this article assumes the AND model.

Deadlock is a significant problem and can be defined as a situation where a group of processes are permanently blocked, as a result of each process having acquired a subset of resources needed for its completion and waiting for the release of the remaining resources held by the other processes in the same group. Finite numbers of resources are available in the system. These resources can be categorized as:

Reusable resources: Defined as the one that can be safely used by only one process at a time and is not depleted by that use, while other is consumable.

Resource: Defined as one that can be created or destroyed. There is no limit on the number of consumable resources of a particular type.

A process must request a resource before using it, and must release the resource after using it. The number of resources requested must not exceed the total number of resources available in the system. A process may utilize a resource in the following sequence:

   i. Request: If the requested resource cannot be granted immediately, then the requesting process must wait until it acquires the desired resources.

  ii. Use: The process can soperate the resource.

 iii. Release: The process releases the resource.

We can represent resource allocation as a graph, where $P \rightarrow R$ means a resource R is currently held by a process P, and process P wants to gain exclusive access to resource R. Deadlock exists when a resource allocation graph has a cycle.

Figure 5.1 illustrates a deadlock condition between 4 processes P_1, P_2, P_3, P_4 and four resources: R_1, R_2, R_3, R_4.

- Process P_1 is holding resource R_1 and wants resource R_2.
- Resource R_3 is held by process P_3 which wants resource R_4.
- Resource R_2 is held by process P_2 which wants resource R_3.
- Resource R_4 is held by process P_4 which wants resource R_1.
- Resource R_1 is held by process P_1 and hence we have deadlock.

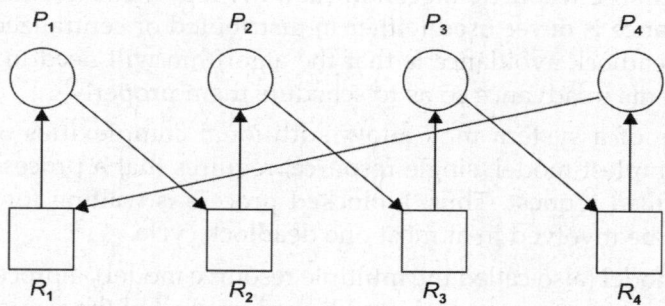

Fig. 5.1: Four deadlocked processes

5.2 System Models

The problem of deadlock has been generally studied in distributed system using the following models:

  i. System has only reusable resources.

  ii. Processes are allowed only exclusive.

 iii. There is only one copy of each resource.

The requirement for a resource differs from time to time. Sometimes a process may require only one resource. But sometimes a process may require more than one resource. It means deadlock models depend on the request made by process. So, depending on the type of request for processes, there are five types of deadlocks:

1. *Single Unit Request Model*: In this there is only one process and a single resource is requested at a time, to execute the process.

2. *AND Request Model*: In this model, if a process needs multiple resources to execute, then, it will execute only in one condition if all the resources are granted to the process. This model is more powerful and allows more concurrency as a process can request for several resources simultaneously.

3. *OR Request Model*: In this model, if a process is granted only some of the resources out of the requested, then also it will be executed.

4. *AND–OR Request Model*: It is a generalization of AND and OR model. In this mode, if any one of the condition of AND or OR Model is satisfied then the process will be executed.

5. *P Out of Q Model*: If a process requests Q resources for execution then out of a Q resources only P resources are available then also the process will be executed. A knot is a sufficient condition for the deadlock.

5.3 Resource versus Communication Model

A deadlock is a condition where a process cannot proceed because it needs to obtain a resource held by another process and it itself is holding a resource that the other process needs.

We can consider two types of deadlock:

i. **Communication deadlock** occurs when process A is trying to send a message to process B, which is trying to send a message to process C, which is trying to send a message to A (Fig. 5.2).

ii. **A resource deadlock** occurs when processes are trying to get exclusive access to devices, files, locks, servers, or other resources (Fig. 5.3).

We will not differentiate between these types of deadlock since we can consider communication channels to be resources without loss of generality.

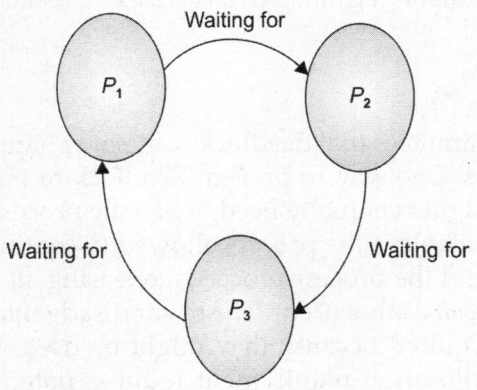

Fig. 5.2: Communication deadlock

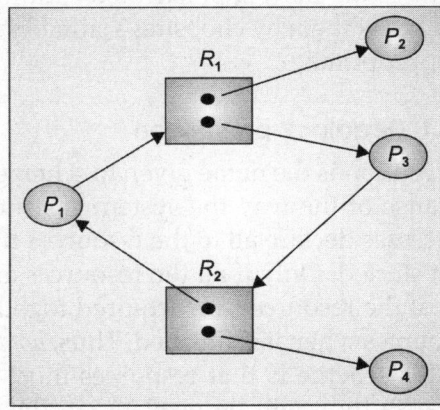

Fig. 5.3: Resource deadlock

In resource deadlock, the processes are simultaneously waiting for several resources and cannot proceed until it acquires all the requested resources, this is known as AND model that process should receive all the requesting resources to make itself executable, where in communication deadlock, the process is waiting for the communication message from different processes but on receiving message from any of the one process it (process waiting for message) can proceed, this is known as OR model, i.e. on receiving message even from a single process it proceeds otherwise the waiting processes are said to be in communication deadlock.

Resource–Allocation Graph: A system resource–allocation graph is a directed graph that is used for describing deadlock the reason for:

- The graph's set of vertices V consists of two partitions: $P = \{P_1, P_2, P_3...,P_n\}$ is the set of active processes, while $R = \{R_1, R_2, R_3,...,R_m\}$ is the set of available resource types

- The graph's set of edge E is also partitioned into two types: request edges ($P_i \rightarrow R_j$) indicate that P_i is waiting for an instance of R_j, while assignment edges ($R_j \rightarrow P_i$) indicate that R_j instance has been allocated to P_i (Resource-Allocation Graphs). By expressing some combination of processes, resource types, requests, and allocations a resource–allocation graph, certain conclusions can be drawn based on the graph's contour:

  - If the graph has no cycles, then there is no deadlock
  - If the graph does have a cycle, then there might be deadlock (i.e. the existence of a cycle is necessary but not sufficient): If the resource types involved in the cycle have only one instance, then we have deadlock. In other words, the existence of a cycle becomes necessary and sufficient when we have only one instance per resource type in the cycle.

5.4 Deadlock Handling Strategies

The problem of deadlocks can be handled in several ways: Prevention, avoidance, and detection. In prevention, some requirements of the system makes deadlocks impossible so that no runtime support is required. Avoidance schemes require decisions by the system while it is running to insure that deadlock will not occur. Detection requires the most sophisticated runtime support, i.e. the system must find deadlocks and break them by choosing a suitable *victim* that is terminated or *aborted* and restarted if appropriate.

5.4.1 Deadlock prevention

Prevention is the name given to schemes, guarantees that deadlocks can never happen because of the way the system is structured. One way to prevent deadlock to make processes declare all to the resources they might eventually need, when the process is first started. Only if all the resources are available is the process allowed to continue. All of the resources are acquired together, and the process proceeds, releasing all the resources when it is finished. Thus, *hold and wait* cannot occur. The major disadvantage of this scheme is that resources must be acquired because they might be used, not because they will be used. Also, the pre-allocation requirement reduces potential concurrency.

Another prevention scheme is to impose an order on the resources and require processes to request resources in increasing order. This prevents *cyclic wait* and thus makes deadlocks impossible. One advantage of prevention is that process aborts are never required due to deadlocks. While most systems can deal with rollbacks, some systems may not be designed to handle them and thus must use deadlock prevention.

5.4.2 Deadlock avoidance

In deadlock avoidance, the system considers resource requests while the processes are running and ensures that those requests do not lead to deadlock. Avoidance based on the *banker's algorithm*, sometimes used in centralized systems, is considered not practical for a distributed system. Two popular avoidance algorithms based on timestamps or priorities are *wound-wait* and *wait-die*. They depend on the assignment of unique global timestamps or priority to each process when it starts. Some authors refer to these as prevention. In wound-wait, if process A requests a resource currently held by process B, their timestamps are compared. B is *wounded* and must restart if it has a larger timestamp (is younger) than A. Process A is allowed to wait if B has the smaller timestamp. Deadlock cycles cannot occur since processes only wait for older processes. In wait-die, if a request from process A conflicts with process B, A will wait if B has the larger timestamp (is younger). If B is the older process, A is not allowed to wait, so it *dies* and restarts. In timeout based avoidance, a process is blocked when it requests a resource that is not currently available. If it has been blocked longer than a timeout period, it is aborted and restarted. Given the uncertainty of message delays in distributed systems, it is difficult to determine good timeout values. These avoidance strategies have the disadvantage that the aborted process may not have been actually involved in a deadlock.

5.4.3 Deadlock detection

Deadlock detection attempts to find and resolve actual deadlocks. These strategies rely on a wait-for-graph (WFG) that in some schemes is explicitly built and analyzed for cycles. In the WFG, the nodes represent processes and the edges represent the blockages or dependencies. Thus, if process A is waiting for a resource held by process B, there is an edge in the WFG from the node for process A to the node for process B. In the AND model (resource model), a cycle in the graph indicates a deadlock. In the OR model, a cycle may not mean a deadlock since any of a set of requested resources may unblock the process. A *knot* in the WFG is needed to declare a deadlock. A knot exists when all nodes that can be reached from some node in a directed graph can also reach that node. In a centralized system, a WFG can be constructed fairly easily. The WFG can be checked for cycles periodically or every time a process is blocked, thus potentially adding a new edge to the WFG. When a cycle is found, a victim is selected and aborted.

5.4.4 Deadlock resolution

Once the deadlock has been detected, it should be resolved to continue proper working of the system. It involves breaking of existing dependencies in the system WFG. It is done by rolling back one or more processes that are deadlocked and assigning their resources to non-blocked processes in deadlock and once resolution has been done then corresponding information should be immediately erased from the system as if it is not cleaned in a timely manner, then it may result in phantom deadlock.

5.5 Control Organizations for Distributed Deadlock Detection

The following three control organizations are common for deadlock detection in distributed systems:

5.5.1 Centralized control

One central site sets up a global WFG and searches for cycles. All decisions are made by the central control node.

- It must maintain the global WFG constantly or
- Periodically reconstruct it.

  The main advantage is that this permits the use of relatively simple algorithms. The disadvantages include the following:
- There is one, single point of failure.
- There can be a communication bottleneck around the site due to all the WFG information messages.
- Furthermore, this traffic is independent of the formation of any deadlock.

5.5.2 Distributed control

In a distributed control organization

- All sites have an equal amount of information.
- All sites make decisions based on local information.
- All sites bear equal responsibility for the final decision in detecting deadlock.
- All sites expend equal effort to the final decision.
- The global WFG is spread across the sites.
- Deadlock detection is initiated whenever a process thinks there might be a problem.
- Several sites can initiate the detection at the same time.

The advantages include the following:

- There is no central point of failure.
- A single node failure cannot cause a crash.
- There is no *one* site with heavy traffic due to the detection algorithm.
- The algorithm is only initiated when process(es) feel there might be a problem.
- The algorithm is not run periodically, only when needed.

The main disadvantage is that resolution may be difficult, as not all sites may be aware of the processes involved in the deadlock, the proof of correctness for this type of algorithm may be difficult.

5.5.3 Hierarchical control

The sites (nodes) are logically connected in a hierarchical structure (such as a tree). A site can detect deadlock in its descendants. This type of algorithm has the best of both the centralized and the distributed deadlock detection algorithms. For efficiency purposes, it is best to keep clusters of interacting processes together in the hierarchy.

5.6 Algorithms for Distributed Deadlock Detection

Centralized deadlock detection

There are two methods to detect deadlock using this model
a. Completely centralized algorithm
b. Ho-Ramamoorthy algorithm

a. **Completely centralized algorithm**: Centralized deadlock detection attempts to imitate the non-distributed algorithm through a central coordinator. Each machine is responsible for maintaining a resource graph for its processes and resources. A central coordinator maintains the utilization graph for the entire system. This graph is the union of the individual graphs. If the coordinator detects a cycle, it finishes off one process to break the deadlock.

In the non-distributed case, all the information on resources usage lies on one system and the graph may be constructed on that system. In the distributed case, the individual sub-graphs have to be propagated to a central coordinator; a message can be sent each time an arc is added or detected. If optimization is needed, a list of added or deleted arcs can be fixed periodically to reduce the overall number of message sent.

On machine A, processor P_0 holds resource S and wants resource R which is held by P_1. The schematic is shown in Fig. 5.4a. Another machine B, has a process P_2, which is holding resource T and wants resource S (Fig. 5.4b). Both of these machines send their messages to the central coordinator (Fig. 5.4c), which maintains the union of the processes A and B. If there is no cycle and hence no deadlock.

There also exist situations where it seems to be deadlock but actually no deadlock occurred is known as *false deadlock* or *phantom deadlock* (Fig. 5.4d).

b. **Ho-Ramamoorthy algorithm:** This algorithm works in two phases:
  i. Two phase algorithm.
 ii. One phase algorithm.

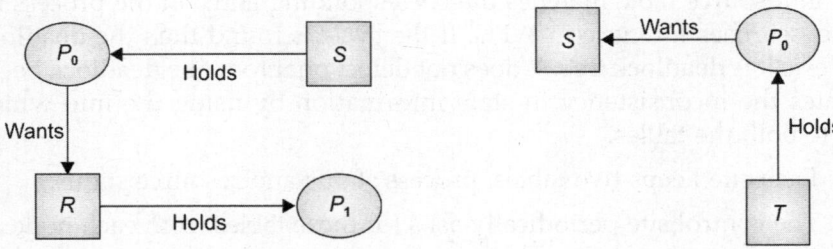

Fig. 5.4a: Resource graph on A **Fig. 5.4b:** Resource graph on B

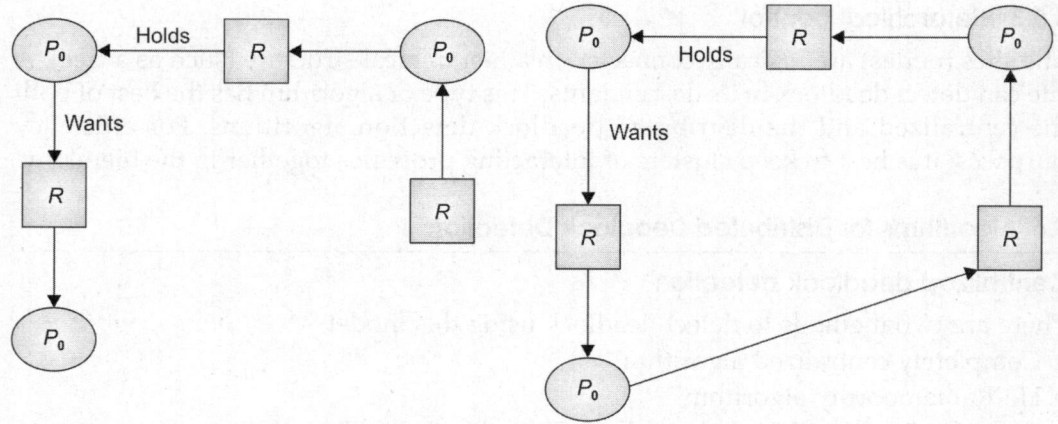

Fig. 5.4c: Resource graph on coordinator   **Fig. 5.4d:** False deadlock

i. **Two-phase algorithm:** In this algorithm every site maintains a status table that contains status of all the processes initiated at that site. The status of a process includes all locked resources and all resources being waited for. A site requests a status table from all sites and then construct the WFG from the information received and search for cycle. If there is no cycle then the system is free from deadlock and if cycle is found then again the status table in updated and whole previous scenario is revised.

- Step 1: Each site has a status table of locked and waited resources.
- Step 2: The control site will periodically ask for this table from each node.
- Step 3: The control node will search the cycle and, if found it will request the table again from each node.
- Step 4: Only the information common in both reports will be analyzed for confirmation of a cycle.

ii. **One-phase algorithm:** This requires only one status report from each site, however, every site maintains two status tables, i.e. a resource status table and a process status table. In the resource status table, site keep the track of transactions that have locked or are waiting for resources which are on hold at that site. The process status table keeps the track of resources locked or waited for, by all the transactions at that site. Every site constructs a WFG using only those transactions for which the entry in resource table matches the corresponding entry in the process table and hence, searches for cycle in WFG. If the cycle is found then the deadlock is detected, else it is deadlock free. It does not detect phantom one deadlock because it eliminates the inconsistency in state information by using the info which is common to both the tables.

- Step 1: Each site keeps two tables, process status and resource status.
- Step 2: The control site periodically asks for these tables from each node.
- Step 3: The control site will then builds and analyze the WFG, looking for cycles and resolving them when found.

Advantage of this algorithm is that, it is faster and requires a few messages as compared to two-phase algorithm while the disadvantage of this algorithm is that it requires more storage spaces as every site maintains two status tables and also exchange large messages as the messages contains both the tables.

5.6.2 Distributed deadlock detection

- All sites completely cooperate to detect a cycle in the state graph that is likely to be distributed over several sites of the system.
- The algorithm can be initiated whenever a process is forced to wait.
- Distributed deadlock can be detected by taking a snapshot of the system and examining it for the condition of a deadlock.
- These algorithms can be divided into four classes:
    i. **Path-pushing:** Path information will sent demo to the waiting node to the blocking node.
    ii. **Edge-chasing:** Probe messages are sent along graph edges.
    iii. **Diffusion computation:** Echo messages are sent along graph edges.
    iv. **Global state detection:** Sweep-out, sweep-in WFG construction and reduction.

i. Path-pushing Algorithm

- In path-pushing algorithms, distributed deadlocks are detected by maintaining an explicit global WFG.
- The basic idea is to build a global WFG for each site of the distributed system.
- In this class of algorithms, at each site whenever deadlock computation is performed, it sends its local WFG to all the neighboring sites.
- After the local data structure of each site is updated, this updated WFG is then passed along to other sites, and the procedure is repeated until some site has a sufficiently complete picture of the global state to announce deadlock or to establish that no deadlocks are present.
- This feature of sending around the paths of global WFG has led to the term path-pushing algorithms.

Obermarck's algorithm

- The site waits for deadlock-related information from other sites.
- The site combines the received information with its local TWF graph to build an updated TWF graph.
- For all cycles '$Ex \to T_1 \to T_2 \to Ex$' which contains the node 'Ex', the site transmits them in string form 'Ex, T_1, T_2, Ex' to all other sites where a sub-transaction of T_2 is waiting to receive a message from the sub-transaction of T_1 at that site.
- For a deadlock, the highest priority transaction detects the deadlock.

Performance of Obermarck's path-pushing algorithm

- $O(n(n-1)/2)$ messages.
- $O(n)$ message sites.
- $O(n)$ delay to detect deadlock.

b. Edge-chasing Algorithm

- In an edge-chasing algorithm, the presence of a cycle in a distributed graph structure has to be verified by propagating special messages called probes, along the edges of the graph.
- These probe messages are different than the request and reply messages.
- The formation of cycle can be deleted by a site if it receives the matching probe sent by it previously.
- Whenever a process that is executing receives a probe message, it discards this message and continues.
- Only blocked processes propagate probe messages along their outgoing edges.
- Main advantage of edge-chasing algorithm is that probes are fixed size messages which are normally very short.

Edge-chasing algorithm (Chandy et al)

- **Sending the probe:**

  if Pi is locally dependent on itself, then deadlock.

  else for all P_j and P_k such that

  (a) P_i is locally dependent upon P_j, and

  (b) P_j is waiting on P_k, and

  (c) P_j and P_k are on different sites, send probe (i,j,k) to the home site of P_k.

- **Receiving the probe:**

  (d) if P_k is blocked, and

  (e) dependent $k(i)$ is false, and

  (f) P_k has not replied to all requests of P_j,

  then begin

  dependent $k(i) :=$ true;

  if $k = i$ then P_i is deadlocked

  else for all P_m and P_n such that

  (a) P_k is locally dependent upon P_m, and

  (b) P_m is waiting on P_n, and

  (c) P_m and P_n are on different sites, send probe (i,m,n) to home site of P_n.
  end.

Performance

- For a deadlock that spans m processes over n sites, $m (n-1)/2$ messages are needed.
- Size of the message 3 words.
- Delay in deadlock detection $O(n)$.

iii. Diffusion computation based algorithm

- In diffusion computation based distributed deadlock detection algorithm; deadlock detection computation is diffused through the WFG of the system.
- These algorithms make use of echo algorithms to detect deadlocks.
- This computation is superimposed on the underlying distributed computation. If this computation terminates, the initiator declares a deadlock.
- To detect a deadlock, a process sends out query messages along all the outgoing edges in the WFG. These queries are successively propagated (i.e. diffused) through the edges of the WFG.
- When a blocked process receives first query message for a particular deadlock detection initiation, it does not send a reply message until it has received a reply message for every query it sent.
- For all subsequent queries for this deadlock detection initiation, it immediately sends back a reply message.
- The initiator of deadlock detection detects a deadlock when it receives reply for every query it has sent out.

Algorithm

Initiation by a blocked process P_i:

send query (i,i,j) to all processes P_j in the dependent set DS_i of P_i;
num (i): = $|DS_i|$; wait$_i(i)$: = true;

Blocked process P_k receiving query (i, j, k):

if this is *engaging* query for process P_k /* first query from P_i */
then send query (i, k, m) to all P_m in DS_k;
num$_k(i)$:= $|DS_k|$; wait$_k(i)$:= true;
else if wait$_k(i)$ then send a reply (i, k, d) to P_j.

Process P_k receiving reply (i,j,k)

if wait$_k(i)$ then
num$_k(i)$: = numk(i) – 1;
if num$_k(i) = 0$ then
if $i = k$ then declare a deadlock.
Else
Send reply (i, k, m) to P_m, which sent the engaging query.

iv. Global state detection algorithm

It is based on two facts of distributed system:
- Take snapshot of distributed WFG.
- Use graph reduction to check deadlock.

Algorithm

Wait$_i$: boolean (:= false)/* records the current status*/
T_i : integer (:= 0)/* current time*/

in (i) : set of nodes whose requests are outstanding at i

out (i) : set of nodes on which i is waiting

P_i : integer (:= 0) /* number of replies required for unblocking */

w_i : real (:= 1.0) /* wait to detect termination of deadlock detection algorithm */

- REQUEST_SEND (i):

  /*executed by node i when it blocks on a p_i-out of q_i request */

  For every node j on which i is blocked do

  out $(i) \leftarrow$ out (i) U {j}; send REQUEST (i) to j;

  set p_i to the number of replies needed; wait$_i$:= true

- REQEST_RECEIVE (j):

  /* executed by node i when it receives a request made by j */

  in $(i) \leftarrow$ in (i) U {j};

- REPLY_SEND (j):

  /* executed by node i when it replies to a request by j */

  in $(i) \leftarrow$ in $(i) - \{j\}$;

  send REPLY (i) to j;

- REPLY_RECEIVE (j):

  /*executed by node i when it receives a reply from j to its request

  if valid reply for the current request then begin

  out$(i) \leftarrow$ out $(i) - \{j\}$; $P_i \leftarrow P_i - 1$;

  if $P_i = 0 \rightarrow$

  　　　　{ wait $i \leftarrow$ false;

  　　　　For all $k \in$ out (i), send CANCEL (i) to k;

  Out $(i) \leftarrow \varphi$}

  end

- CANCEL_RECEIVE (j):

  /* executed by node i when it receives a cancel from j */

  if $j \in$ in (i) then in

  $(i) \leftarrow$ in $(i) - \{j\}$;

5.6.3 Hierarchical deadlock detection

- Follows Ho-Ramamoorthy's one-phase algorithm. More than one control site organized in hierarchical manner.
- Each control site applies one-phase algorithm to detect (intra-cluster) deadlocks.
- Central site collects info from control sites, applies one-phase algorithm to detect intra-cluster deadlocks.

5.7 Graph Reduction Method

General idea: To simulate the result of execution, assuming all unblocked processes complete without requesting any more resources.

- While there is an unblocked process, remove the process and all (resource-holding) edges to it.
- There is deadlock if the remaining graph is non-null.

A reduction is most optimistic set of operations that unblocks one or more blocked processes.

Reducing general resource is a resource graph by an unblocked process P_i.

1. For each reusable resource R_i
   a. Delete all assignment and request edges.
   b. For each assignment for a reusable resource increament the number of available unit (r_j)
2. For each consumable resource (r_j)
   a. Decrement the available unit (r_j) by number of request edges.
   b. Delete all request and producer edge.
   c. If P_i is producer of (R_j), then the subset of available unit (r_j) for that resource to ∞.

Deadlock is prevented when the graph has cycles.

SUMMARY

The deadlock is a process to block the waiting on a condition that will never satisfy. There are various conditions in deadlock like hold wait. Mutual exclusion, no preemption and circular wait. The various strategies for handling deadlock are deadlock prevention, avoidance and detection. The various types of system models are also provided like AND, single, OR, Pout of Q on the basis of processors. The deadlock can be detected by using various algorithms such as path pushing, edge chasing, diffusion computation and global state detection. Through graph reduction method a set of operations unblocks one or more blocked processes.

REVIEW QUESTIONS

1. State the differences between resource and communication deadlock.
2. What are phantom deadlock? Explain the algorithm which can detect phantom deadlock?
3. Explain Obermarck's algorithm of path pushing.
4. Explain various deadlock handling strategies in distributed system.
5. What are the shortcomings of Ho-Ramamoorthy's two-phase algorithm for deadlock detection?
6. Explain the control organization for distributed deadlock detection.

UNIT 3

Chapter 6: Agreement Protocols

Chapter 7: Distributed Resource Management

6

Agreement Protocols

Contents

6.1 Introduction

As we know in distributed systems (DS) many sites are involved at a time to complete a single task, i.e. to achieve a common goal, so it is required that each site should agree to reach a mutual agreement to depict whether the goal is reached or not by them. Hence, to reach an agreement all sites should have the knowledge about the values of other sites. If all the sites agree on a common/same value then and only then agreement has reached *else if* the values are conflicting, i.e. not same then, it may be possible that system contains some faulty processors due to which problem occurs in reaching an agreement. Hence, it becomes difficult to reach to an agreement in case of a system prone to failure.

- Processes/sites in DS often compete as well as cooperate to achieve a common goal.

- Mutual trust/agreement is very much required.

- In distributed databases, there may be a situation where data managers have to decide whether to commit or abort a transaction.

- When there is no failure, reaching an agreement is easy.

- However, in case of failures, processes must exchange their values with other processes and relay, the values received from others several times to isolate the effect of faulty processor.
- Agreement protocols help to reach an agreement in presence of failures.

In distributed systems, where sites (or processors) often compete as well as cooperate to achieve a common goal, it is often required that sites reach mutual agreement.

Example: In distributed database systems, data managers at sites must agree on whether to commit or abort a transaction.

The formal setting for a distributed agreement protocol is the following: There are M processors, $P = P_1,...,P_M$ that are trying to reach agreement. A subset F of the processors are faulty, and remaining processors are non-faulty. Each processor $P_i \in P$ stores a value V_i. During the agreement protocol, the processors calculate an agreement value A_i. After the protocol ends, the following two conditions should hold:

i. For every pair p_i and p_j of non faulty processors, $A_i = A_j$. This value is the agreement value.

ii. The agreement value is a function of the initial values $\{V_i\}$ of the non faulty processors $(P-F)$.

6.2 System Models

Agreement problems have been studied under the following system models:

i. There are n processors in the system and at most m of the processors can be faulty.

ii. The processors can directly communicate with other processors by message exchanging. Thus, the system is logically fully connected.

iii. A receiver (processor) always knows the identity of sender (processor) of the message.

iv. The communication medium is reliable, means it delivers all messages without introducing any occur.

v. The processors are proving to failure.

6.2.1 Synchronous versus asynchronous computation

Synchronous computation:

i. Processes run in lock step manner [process receives a message sent to it earlier, performs computation and sends a message to other process].

ii. Step of synchronous computation is called round.

iii. In each step/round I, processes receive messages that were sent in the previous step/round $i-1$.

iv. Processes then do some computation and send out messages, that will be received in step/round $i+1$.

v. Message delays or a slow processor slows down the whole system.

vi. Agreement protocols to be discussed assume synchronous computation.

Asynchronous computation:

i. Computation does not proceed in lock step.

ii. Process can send receive messages and perform computation at any time.

6.2.2 Mode of processor failures

Processor can fail in three modes:
  i. *Crash fault*:
      i. Processor stops and never resumes operation.
  ii. *Omission fault*:
      a. Processor omits to send message to some processors.
  iii. *Malicious fault*:
      a. Also known as Byzantine faults.
      b. Processor may send fictitious values/message to other processes to confuse them
      c. Tough to detect/correct.

6.2.3 Authenticated versus non-authenticated messages

The messages can be classified as:
1. Authenticated messages
2. Non-authenticated messages
1. *Authenticated messages*
    a. Also known as signed message
    b. Processor cannot forge/change a received message
    c. Processor can verify the authenticity of the message
    d. It is easier to reach on an agreement in this case
2. *Non-authenticated messages*
    a. Also known as oral message
    b. Processor can forge/change a received message and claims to have received it from others
    c. Processor cannot verify the authenticity of the message in this case

6.2.4 Performance aspects of agreement protocols

* Following metrics are used:
    i. Time: Number of rounds needed to reach an agreement
    ii. Message traffic: Number of messages exchanged to reach an agreement.
    iii. Storage overhead: Amount of information that needs to be stored at processors during execution of the protocol.

6.3 Classification of Agreement Models

There are three well-known agreement problems in distributed systems:
  1. **Byzantine agreement problem:**
      * A single value is to be agreed upon.
      * Agreed value is initialized by an arbitrary processor and all non faulty processors have to agree on that value.
  2. **Consensus problem:**
      * Every processor has its own initial value and all non faulty processors must agree on a single common value.

3. Interactive consistency problem:
- Every processor has its own initial value and all non faulty processors must agree on a set of common values
- In all the previous mentioned problems, all non faulty processors must reach an agreement
- In all the previous mentioned problems, all non faulty processors must reach an agreement
- In Byzantine and consensus problems, agreement is on a single value
- In interactive consistency problem, agreement is on a set of common values
- In Byzantine agreement problem, only one processor initializes the value where as in other two cases, every processor has its own initial value

6.3.1 Byzantine agreement problem

- Source processor [any arbitrarily chosen processor] broadcasts its values to others (Fig. 6.1).
- Solution must meet the following objectives:
  1. Agreement: All non faulty processors agree on the same value.
  2. Validity: If source is non-faulty, then the common agreed value must be the value supplied by the source processor.
     - If source is faulty then all non faulty processors can agree on any common value.
     - Values agreed upon by faulty processors is irrelevant

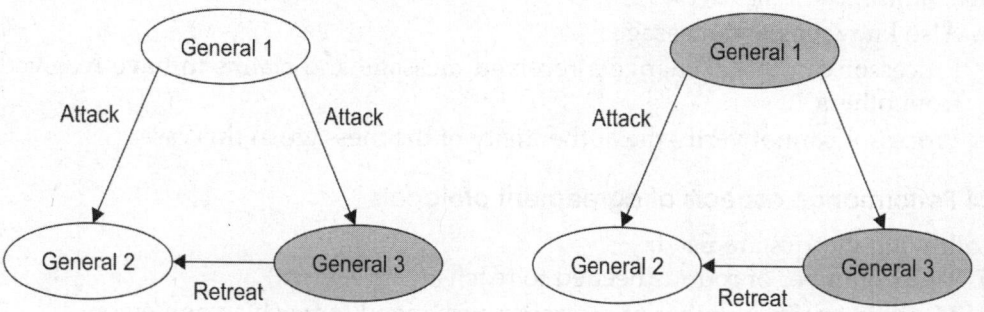

Fig. 6.1: Example of Byzantine agreement problem

Example: The problem shown in Fig. 6.1 is generally known as Byzantine generals problem.

Byzantine general problem

1. The classic problem
   - Each division of Byzantine army directs its own general
   - Generals, some of which are traitors, communicate each other by messengers
   - Requirements:
     i. All loyal generals decide upon the same plan of action
     ii. A small number of traitors cannot cause the loyal generals to adopt a bad plan

2. The problem can be restated as:
   - All loyal generals receive the same information upon which they will somehow get to the same decision
   - The information sent by a loyal general should be used by all the other loyal generals
3. The above problem can be reduced into a series of one commanding general and multiple lieutenants problem (Byzantine generals problem):
   - All loyal lieutenants obey the same order
   - If the commanding general is loyal, then every loyal lieutenant obeys the order it sends.

Reliability by majority voting

- One way to achieve reliability is to have multiple replica of system (or component) and take the majority voting among them
- In order to have the majority voting and yield a reliable system, the following two conditions should be satisfied:
  - i. All non faulty components must use the same input value
  - ii. If the input unit is non-faulty, then all non faulty components use the value it provides as input

6.3.2 Consensus problem

- Every processor broadcasts its initial value to all other processors.
- Initial values may be different for different processors.
- Protocol must meet following objectives:
  1. Agreement: All non faulty processors agree on the same single value.
  2. Validity: If initial value of every non faulty processor is v, then the common agreed value by all non faulty processors must be v.
     - If initial value of non faulty processors is different then all non faulty processors can agree on any common value.
     - Values agreed upon by faulty processors is irrelevant

6.3.3 Interactive consistency problem

- Every processor broadcasts its initial value to all other processors.
- Initial values may be different for different processors.
- Protocol must meet the following objectives:
  1. Agreement: All non faulty processors agree on the same vector $(v_1, v_2, .., v_n)$.
  2. Validity: If ith processor is non faulty and its initial value is v_i, then the ith value agreed by all non faulty processors must be v_i.
     - If jth processor is faulty then all non faulty processors can agree on any common value v_j
     - Value agreed upon by faulty processors is irrelevant

6.4 Relations Among Agreement Problems

All three agreement problems are closely related, e.g. the Byzantine agreement problem is a special case of the interactive consistency problem, in which the initial value of only one processor is of interest. Conversely, if each of the processors runs a copy of the Byzantine agreement protocol, the interactive consistency problem is solved. Likewise, the consensus problem can be solved using the solution of the interactive consistency problem. This is because all non faulty processors can compute the value that is to be agreed upon by taking the majority value of the common vector which can computed by an interactive consistency protocol, or by choosing a default value if a majority does not exist.

Thus, solutions to the interactive consistency and consensus problems can be derived from the solutions to the Byzantine agreement problem. In other words, the Byzantine agreement problem is primitive to the other two agreement problems. For this reason, we will focus solely on the Byzantine agreement problem for the rest of the chapter.

However, it should by no means be concluded that the Byzantine agreement problem is weaker than the interactive consistency problem or that the interactive consistency problem is weaker than the consensus problem. In fact, there is no linear ordering of this sort among these agreement problems. For example, the Byzantine agreement problem can be solved using a solution to the consensus problem in the following manner:

1. The source processor sends its value to all other processors, including itself.
2. All the processors, including the source, run an algorithm for the consensus problem using the values received in the first step as their initial values.

The above two steps solve the Byzantine agreement problem because if the source processor is non-faulty, then all the processors will receive the same value in the first step and all non faulty processors will agree on that value as a result of the consensus algorithm in the second step, and if the source processor is faulty, then the other processors may not receive the same value in the first step, however, all non faulty processors will agree on the same value in the second step as a result of the consensus algorithm. Thus, the Byzantine agreement is reached in both cases. However, the $n - 1$ extra messages are sent at the first step.

6.5 Solution to Byzantine Agreement Problem

- The solution of this problem was first defined and solved by Lamport *et al*, its solution was also given by Lamport first under the situation of processor failure.
- According to the concept of Byzantine agreement problem source processor is taken to broadcast its initial value to all other processors in the system. The source processor is chosen randomly and no sequential ordering is used to choose a source processor.
- Processors send their values to other processors and also relay received values to others.
- During execution faulty processors may confuse by sending conflicting values. However, if faulty processors dominate in number, they can prevent non faulty processors from reaching an agreement.

- Number of faulty processors should not exceed certain limit, shown by Pease, that in a fully connected network, it is impossible to reach an agreement if number of faulty processors 'm' exceeds $(n-1)/3$ where, n = number of processors.

Hence, the solution of Byzantine agreement problem is based on the presence of number of faulty and non faulty processors.

6.6 Lamport-Sheoshtak-Pease Algorithm

Lamport *et al* proposed an algorithm for solving the Byzantine agreement problem. This algorithm is also known as "Lamport–Sheostak–Pease algorithm". They proposed the algorithm for oral messages (i.e. non-authenticated messages), so it is also referred to as "oral message algorithm"(OM).

The algorithm works if the following is true:

$$n \geq = 3m + 1$$

where n = number of processors, and

m = number of faulty processors

The algorithm works differently with or without faulty processors and hence can be defined recursively as follows:

- **Algorithm OM(0), i.e. $m = 0$**

Step 1: Source processor sends its values to every processor

Step 2: Each processor uses the value it receives from source. [If no value is received default value 0 is used]

- **Algorithm OM(m), i.e. $m > 0$**

Step 1: The source processor sends its value to every processor.

Step 2: For each i, let v_i be the value processor i receives from source [default value 0 if no value received]

Step 3: Processor I acts as the new source and initiates algorithm OM $(m-1)$ where it sends the value v_i to each of the $n-2$ other processors.

Step 4: For each i and j (not i), let v_j be the value processor i received from processor j in step 3. Processor I uses the value majority $(v_1, v_2...,v_n-1)$. The function majority $(v_1, v_2...,v_{n-1})$ computes the majority value if exists otherwise it uses default value 0.

6.6.1 Complexity of algorithm

1. The algorithm OM (M) includes $(n-1)$ separate execution of algorithm OM $(M-1)$ each of which invokes $(n-2)$ execution of algorithm OM $(M-2)$ and so on.

2. So, there will be $(n-1)$ $(n-2)$ $(n-3)$,..., $(n-k)$ separate execution of algorithm OM,..., will be $(m-k)$, $k-1$, 2, 3, $m+1$.

3. Hence, the message complexity is $O(nm)$ and time complexity is $m+1$ rounds.

Example for algorithm:

$$n = 4$$
$$M = 1$$

This proves that we will reach to the result.

Demonstration of Example:

1. Let there are four processors P_0, P_1, P_2, P_3 which communicates with each other having values 0 and 1. P_0 is source processor and P_2 is faulty.

2. Processor P_0 starts algorithm OM (1) by initiating its value as 1 and sends it to all other processors (Fig. 6.2).

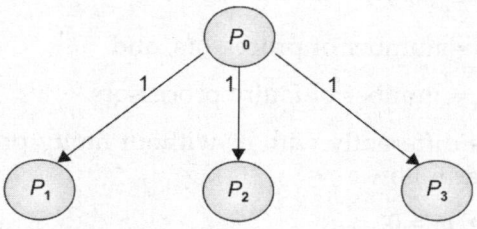

Fig. 6.2

3. Now we execute step 2 of algorithm OM (m), $m = 1$.

4. After receiving value 1 from source processor (P_0). P_1, P_2, P_3 processors execute algorithm OM (0) (Fig. 6.3).

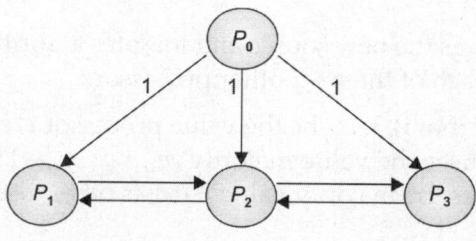

Fig. 6.3

5. The processors P_1 and P_3 are non faulty and send value 1 to (P_2, P_3) and (P_1, P_2), respectively.

6. P_2 sends 1 to P_1 and 0 to P_3 as it is faulty.

7. Next, processor P_1, P_2, P_3 execute step 4 of OM (1) for deciding the majority value.

8.

Processor	Vector received	Majority value
P_1	$(1, 1, 1)$	1
P_2	$(1, 1, 1)$	1
P_3	$(1, 1, 0)$	1

9. Hence, all the 3 processors agree to a same value, i.e. 1 and Byzantine agreement problem is solved.

6.7 Solution to Impossible Results

Now, we will show that the solution to Byzantine problem cannot be reached in three processors if one of them is faulty. Sometimes, in agreement problem we reached to a situation when it is quite impossible to solve the problem and this is known as impossible result and such situation is found in a system with more than two processors.

We consider three processors P_1, P_2 and P_3 which communicates with each other. The two possibilities are:

1. Source processor is faulty.
2. Source processor is non-faulty.

Case 1: The source processor P_1 is faulty (Fig. 6.4).

Hence, we come to know that P_2 receives 1 from P_1 hence sends 1 to P_3 and P_3 receives 0 from P_1 so, its sends 0 to P_3 so, P_2 is forced to agree on the value 1 received from source processor and P_3 will agree on value 0, hence, the agreement cannot be reached and also it does not satisfy the condition:

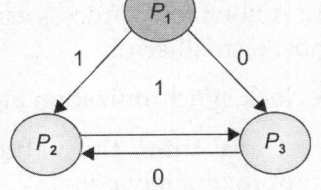

Fig. 6.4: P_1 is faulty

$$n \geq 3m + 1 \qquad \because \quad n = 3$$

$$3 \geq 3\,(1) + 1$$

$$3 \geq 4 \quad \therefore \text{ false}$$

So, result for agreement problem is impossible.

Case 2: P_1, the source processor is non faulty and P_3 is faulty.

P_1 is the source processor so, it initiates its value as 1 and sends it to P_2 and P_3. Now, after receiving value from P_1, P_2 sends the same value to P_3 because it is non faulty and P_3 is faulty (Fig. 6.5) so, it changes the value and send 0 to P_2.

Now, to decide the agreement value, P_1 will give 1, P_2 is forced to give 1, hence both the non faulty processor agrees on the same value. So, we can say that agreement has reached, although it also does not satisfy the condition of algorithm provided for the solution of agreement problem by Lamport *et al.*

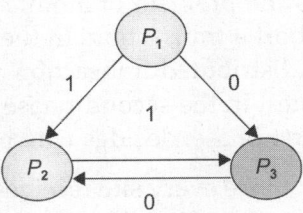

Fig. 6.5: P_3 is faulty

i.e. $\qquad\qquad n \geq 3m + 1$

6.8 Applications of Agreement Algorithms

The agreement algorithms are very useful in the conditions where processors will reach an agreement regarding their values in the presence of malicious failures. Two such applications are:

1. Clock synchronization in distributed database.

2. Atomic commit in distributed database.

6.8.1 Clock synchronization (Fault tolerant)

In distributed systems, it is often necessary that sites (or processes) maintain physical clocks that are synchronized with one another. Since physical clocks have a drift problem, they must be periodically resynchronized. Such periodic synchronization becomes extremely difficult if the Byzantine failures are allowed. This is because a faulty process can report different clock values to different processes. The description of clock synchronization in this section is based on the work of Lamport and Melliar–Smith. We make the following assumptions regarding the system:

A_1: All clocks are initially synchronized to approximately the same values.

A_2: A non faulty process clock runs at approximately the correct rate (The correct rate means one second of clock time per second of real time. No assumption is made about a faulty clock).

A_3: A non faulty process can read the clock value of another non faulty process with at most a small error.

A clock synchronization algorithm should satisfy the following two conditions:

- At any time, the values of the clocks of all non faulty processes must be approximately equal.

- There is a small bound on the amount by which the clock of a non faulty process is changed during each resynchronization.

The latter condition implies that resynchronization does not cause a clock value to jump arbitrarily far, thereby preventing the clock rate from being too far from the real time.

6.8.2 Atomic Commit in Distributed Database

In the problem of atomic commit, sites of a DDBS must agree whether to commit or abort a transaction. In the first phase of the atomic commit, sites execute their part of a distributed transaction and broadcast their decisions (commit or abort) to all other sites. In the second phase, each site, based on what it received from other sites in the first phase, decides whether to commit or abort its part of the distributed transaction.

Since every site receives a identical response from all other sites, they will reach the same decision. However, if some sites behave maliciously, they can send a conflicting response to other sites, causing them to make conflicting decisions.

In these situations, we can use algorithms for the Byzantine agreement to insure that all non faulty processors reach a common decision about a distributed transaction. It works as follows:

i. In the first phase, after a site has made a decision, it starts the Byzantine agreement.

ii. In the second phase, processors determine a common decision based on the agreed vector of values.

Two-phase commit:

The two-phase commit protocol has following two phases (Fig. 6.6)

i. Voting phase

ii. Decision phase

The coordinator sends vote request message to all the participating processors. On receiving this message every site sends vote-commit or vote-abort message to the coordinator. On collecting all the messages/votes it decides, if all the participants have sent commit then coordinator will send global-commit message to all the participants and they commit all local transactions and if all the participants send abort message then, the coordinator will send global-abort message to the participants and they will abort their local transactions.

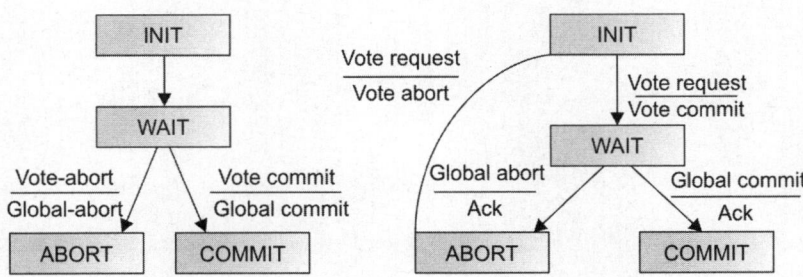

Fig. 6.6: Two-phase commit protocol

SUMMARY

In distributed system, there is a common goal to reach the mutual agreement. There are agreement protocols in which faulty processor and non faulty processor must exchange their values. There are various system models and various types of computation method, and these are various agreement problems like Byzantine agreement problem, consensus problem, etc. There is a solution to Byzantine agreement problem given to solve the examples. There are some application through which agreement algorithms like as clock synchronization and atomic commit in database system can be achieved.

REVIEW QUESTIONS

1. What do you understand by system models?
2. How the agreement protocols are classified? Explain in detail.
3. Describe the complexity of Lamport's algorithm for Byzantine agreement problem.
4. Describe in brief the atomic commit in database with a suitable example.
5. Show the Byzantine agreement cannot always be reached among four processors, if two of them are faulty.
6. Write a short notes on agreement protocols.
7. Explain the algorithm for the solution of Byzantine agreement problems.

7

Distributed Resource Management

Contents

7.1 Introduction

The distributed file system works as the resource management component which manages the availability of files among various sites. It maintains the network transparency. It means, the availability of file distributed over the network in the same way the file residing at one location. So, the user can easily access the file irrespective of their location. Hence, it implements a common file system that can be shared by all the autonomous components (systems).

- A file is a sequence of bytes which have attributes, which are information about the file such as the owner, size, creation data, and access permissions.

  - In UNIX, these attributes and the file addresses are saved in the *i* node.

- A file service usually provides primitives to read and write some of the attributes and the data.

  - The upload/download model: Files moved to clients. Clients do all the operations and send it back.

  - The remote access model: Files stay on the server, and the server do all the modifications.

- In a distributed file system, files can be stored at any machine.
- For higher performance, several machines, referred to file servers, are dedicated to storing files and performing storage and retrieve operations.
- Name server: A process that maps names specified by clients to stored objects such as files and directories.
- Cache manager: A process that implements file caching.
  - Can be present at both clients and servers.

7.1.1 File usage characteristics

- Most files are small (less than 10 kb).
- Reading is much more common than writing.
- Reads and writes are sequential, random access is rare.
- Most files have a short lifetime.
- File sharing is unusual.
- The average process uses only a few file.

7.1.2 Goals of distributed file system

The goals of distributed file system are as follows:

  i. Network transparency (access transparency)

  ii. Availability

Network transparency (access transparency)

- Users should be able to access files over a network easily, as if the files were stored locally.
- Users should not have to know the physical location of a file to access it.
- Transparency can be addressed through naming and file mounting mechanism
- Location transparency: Filename does not specify physical location.
- Location independence: Files can be moved to new physical location, no need to change references to them.
- Location independence → location transparency, but the reverse is not necessarily true.

Availability

- Files should be easily and quickly accessible.
- The number of users, system failures, or other consequences of distribution should not compromise the availability.
- Addressed mainly through replication.

7.2 Architecture of Distributed File System

1. In distributed system files can be stored at any machine computation can be performed at many machines. However, for higher performance, several machines, referred to as file servers, are dedicated to storing files and performing storage and retrieval operations (Figs 7.1 and 7.2).

2. All remaining machine/computer are used for computational purposes. These machines are referred to as clients and they access the files stored on servers. Some clients contain local hard disk for storage purpose.

3. Services provided by the distributed file system:

   i. Name server: Maps names specified by clients to stored objects such as files and directories. The mapping occurs when a process references a file or directory for the first time.

   ii. Cache server: Implements the file caching in this copy of data stored at a remote file server. It is brought to client machine whenever required. Subsequent access to the data is performed locally at the client, thereby reducing the access delays due to network latency.

4. Cache managers at the servers caches files in the memory to reduce delays due to disk latency.

5. There is problem of inconsistency if multiple clients cache the file and modify it. To avoid these problem cache managers of client and server coordinates to perform data storage and retrieval operations.

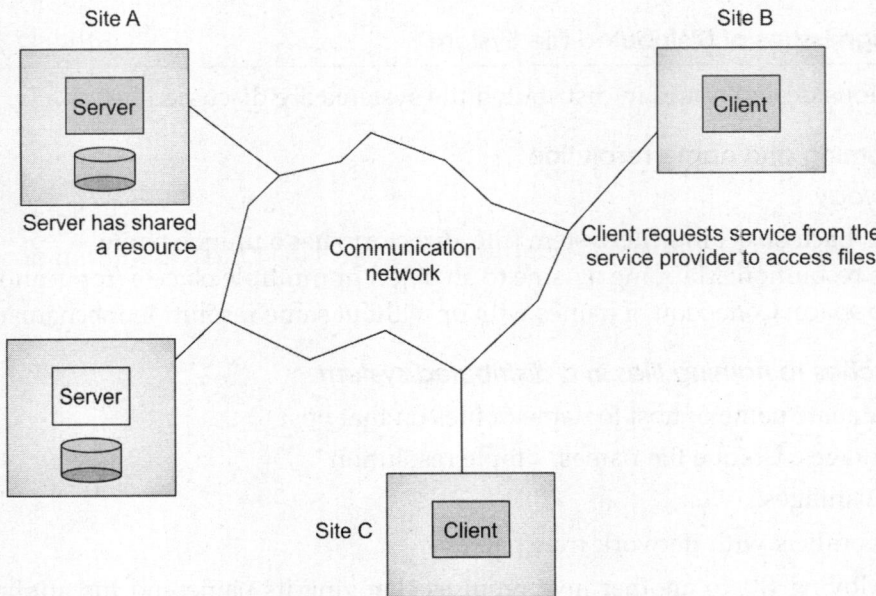

Fig. 7.1: Architecture of distributed file system

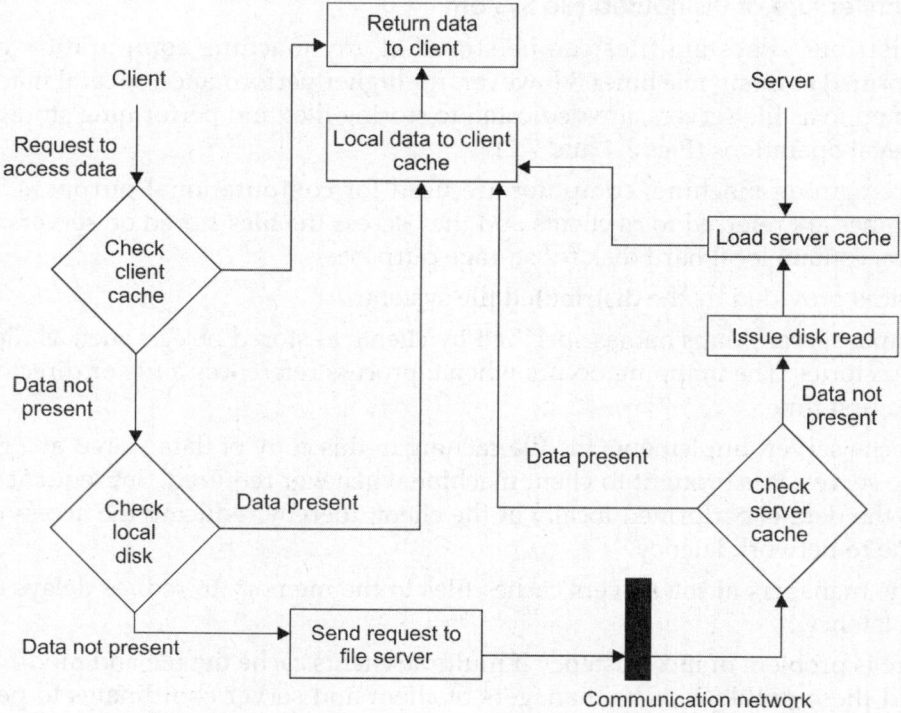

Fig. 7.2: Flow chart showing data access actions

7.3 Design Issues of Distributed File System

The various design issues in distributed file systems are discussed below.

7.3.1 Naming and name resolution

Terminology

a. Name: Each object in a file system (file, directory) has a unique name
b. Name resolution: Mapping a name to an object or multiple objects (replication)
c. Name space: Collection of names with or without same resolution mechanism

Approaches to naming files in a distributed system

a. Concatenate name of host to name of files on that host

   Advantages: Unique file names, simple resolution

   Disadvantages:

     i. Conflicts with network transparency

     ii. Moving file to another host requires changing its name and the applications using it.

b. Mount remote directories onto local directories

     i. Requires that host of remote directory is known

ii. After mounting, files referenced location should be transparent (i.e. file name does not reveal its location)

c. Have a single global directory

i. All files belong to a single name space

ii. Limitation: Having unique system, wide file names require a single computing facility or cooperating facilities

Contexts

a. Solve the problem of system, i.e. wide unique names, by partitioning a name space into contexts (geographical, organizational, etc.)

b. Name resolution is done within that context

c. Interpretation may lead to another context

File name = context + name local to context

Name server

a. Process that maps filenames to objects (files, directories)

b. Implementation options

i. Single name server

Simple implementation, reliability and performance issues

ii. Several name servers (on different hosts)

- Each server responsible for a domain
- Example:
  - Client requests access to file '*A/B/C*'
  - Local name server looks up a table (in kernel)
  - Local name server points to a remote server for '*A/B/C*' mapping

7.3.2 Caches on disk or main memory

Caching at the client: Main memory versus disk

a. Main memory: (+) fast, (+) works for diskless clients, (–) expensive memory, (–) complex virtual memory management.

b. Disk: (+) large files, (+) simpler virtual memory management (–) requires local disk.

Cache consistency

a. Server initiated

i. Server informs cache managers when data in client caches is stale

ii. Client cache managers invalidate stale data or retrieve new data

iii. Disadvantage: Extensive communication

b. Client initiated

i. Cache managers at the clients validate data with server before returning it to clients

ii. Disadvantage: Extensive communication

c. Prohibit file caching when concurrent–writing

    i. Several clients open a file, at least one of them for writing

    ii. Server informs all clients to purge that cached file

d. Lock files when concurrent–write sharing (at least one client opens for write)

Writing policy

a. Once a client writes into a file (and the local cache), when should the modified cache be sent to the server.

b. Options:

    i. Write-through: All writes at the clients, immediately transferred to the servers

       – Advantage: Reliability.

       – Disadvantage: Performance, it does not take advantage of the cache.

    ii. Delayed writing: Delay transfer to servers

       – *Advantages*:

          • Many writes take place (including intermediate results) before a transfer.

          • Some data may be deleted.

       – *Disadvantage*: Reliability issue.

    iii. Delayed writing until file is closed at client

       – For short open intervals, same as delayed writing.

       – For long intervals, reliability issue.

Availability

a. Issue: What is the level of availability of files in a distributed file system?

b. Resolution: Use replication to increase availability, i.e. many copies (replicas) of files are maintained at different sites / servers

c. Replication issues:

    i. How to keep replicas consistent

    ii. How to detect inconsistency among replicas

d. Unit of replication

    i. File

    ii. Group of files

    I. Volume: Group of all files of a user or group or all files in a server

       – Advantage: Ease for implementation

       – Disadvantage: Wasteful, user may need only a subset replicated

    II. Primary pack versus pack

       – Primary pack: All files of a user

       – Pack: Subset of primary pack. Can receive a different degree of replication for each pack

Scalability

a. Issue: Can the design support a growing system

b. Example: Server-initiated cache invalidation complexity and load, increased with size of system.

Possible solutions:

   i. Do not provide cache invalidation service for read-only files.

   ii. Provide design to allow users to share cached data.

c. Design files servers for scalability: Threads, SMPs, clusters.

Semantics

a. Expected semantics: A read will return data stored by the latest write

b. Possible options:

   i. All read and writes go through the server

   • Disadvantage: Communication overhead.

   ii. Use of lock mechanism

   • Disadvantage: File not always available.

7.4 Mechanism for Building Distributed File System

7.4.1 Mounting

a. The mount mechanism binds together several file name spaces (collection of files and directories) into a single hierarchically structured name space (Fig. 7.3) (Example: UNIX and its derivatives).

b. A name space 'A' can be mounted (bounded) at an internal node (mount point) of a name space 'B'.

c. Implementation: Kernel maintains the *mount table,* mapping mount points to storage devices.

d. Location of mount information:

   i. Mount information maintained at clients:

   Each client mounts every file system.

   Different clients may not see the same filename space.

   If files move to another server, every client needs to update its mount table.

   Example: SUN NFS.

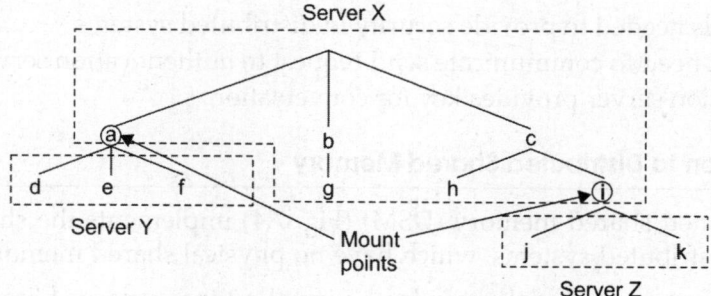

Fig. 7.3: Name space hierarchy

ii. Mount information maintained at servers:

Every client see the same filename space.

If files move to another server, mount info at server only needs to change.

Example: Sprite file system.

7.4.2 Caching

a. Improves file system performance by exploiting the locality of reference.

b. When client references a remote file, the file is cached in the main memory of the server (server cache) and at the client (client cache).

c. When multiple clients modify shared (cached) data, cache consistency becomes a problem.

d. It is very difficult to implement a solution that guarantees consistency.

7.4.3 Hints

a. Treat the cached data as hints, i.e. cached data may not be completely accurate.

b. Can be used by applications that can discover that the cached data is invalid and can recover.

Example:

- After the name of a file is mapped to an address, that address is stored as a hint in the cache.
- If the address later fails, it is purged from the cache.
- The name server is consulted to provide the actual location of the file and the cache is updated.

7.4.4 Bulk data transfer

a. Observations:

i. Overhead introduced by protocols does not depend on the amount of data transferred in one transaction

ii. Most files are accessed in their entirety

b. Common practice: When client requests one block of data, multiple consecutive blocks are transferred

7.4.5 Encryption

a. Encryption is needed to provide security in distributed systems

b. Entities that need to communicate send request to authentication server

c. Authentication server provides key for conversation.

7.5 Introduction to Distributed Shared Memory

- The distributed shared memory (DSM) (Fig. 7.4) implements the shared memory model in distributed systems, which have no physical shared memory

- The shared memory model provides a virtual address space shared between all nodes

- The overcome the high cost of communication in distributed systems, DSM systems move data to the location of access.

- Data moves between main memory and secondary memory (within a node) and between main memories of different nodes.

- Each data object is owned by a node, initial owner is the node that created object and ownership can change as object moves from node to node.

- When a process accesses data in the shared address space, the mapping manager maps shared memory address to physical memory (local or remote).

- DSM allows programs that used to operate on same computer to be easily adapted to operate on separate computers and also the program running on separate computers to share data without programmer having to deal with sending messages.

- Howerver, DSM is not suitable for all architectures. Client-server systems are generally less suited for DSM, but a server may be used to assist in providing DSM functionality for data shared between clients (Fig. 7.4).

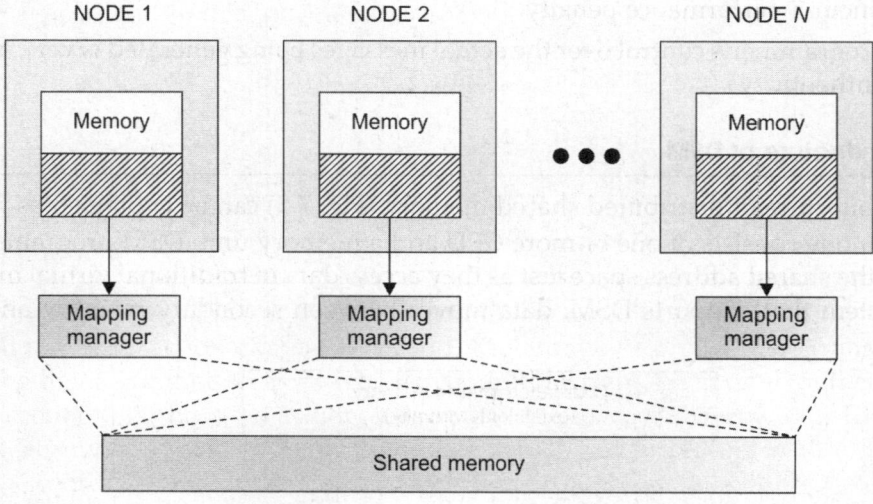

Fig. 7.4: Distributed shared memory

7.6 Advantages and Disadvantages of DSM

7.6.1 Advantages

1. Data sharing is implicit, hiding data movement (as opposed to 'send'/'receive' in message passing model.

2. Passing data structures containing pointers is easier (in message passing model, data moves between different address spaces)

3. Moving entire object to user takes advantage of locality difference

4. Less expensive to build than the tightly coupled multiprocessor system: Off-the-shelf hardware, no expensive interface to shared physical memory

5. Very large total physical memory for all nodes: Large programs can run more efficiently

6. No serial access to common bus for shared physical memory like in multiprocessor systems

7. Programs written for shared memory multiprocessors can be run on DSM systems with minimum changes

7.6.2 Disadvantages

1. Programmers need to understand consistency models, to write correct programs

2. DSM implementations use asynchronous message-passing, and hence cannot be more e_client than message-passing implementations

3. By yielding control to DSM manager software, programmers cannot use their own message-passing solutions.

4. Must provide for protection against simultaneous access to shared data objects (locks, etc.).

5. May incur a performance penalty.

6. The programmer's control over the actual messages being generated is very minimal (i.e. authenticity).

7.7 Architecture of DSM

The architecture of distributed shared memory (Fig. 7.5) can be explained as follows:
Each node consists of one or more CPU and a memory unit. DSM programs access data in the shared address space just as they access data in traditional virtual memory. In a system that supports DSM, data moves between secondary memory and main

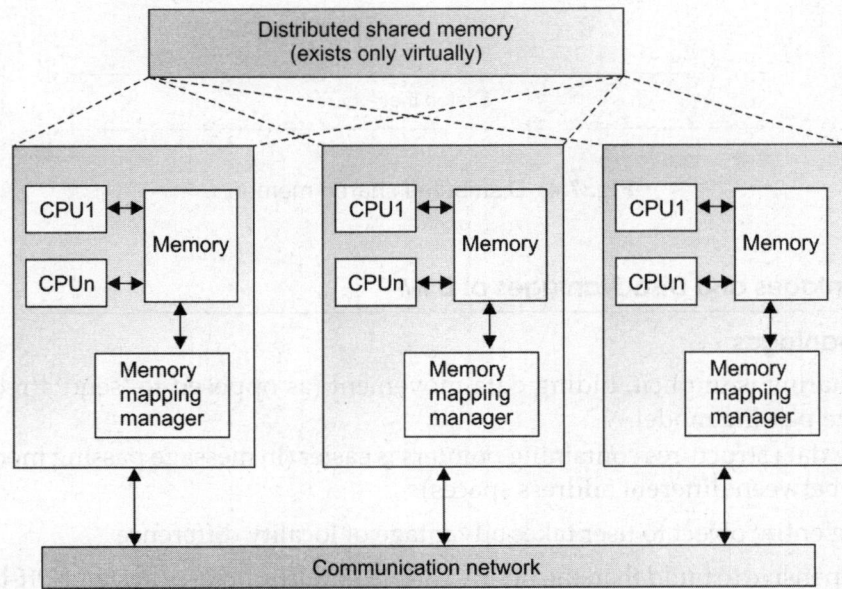

Fig.7.5: Architecture of DSM

memory as well as between main memories of different nodes. Each node can own data stored in the shared address space, and ownership can change when data moves from one node to another. When a process access data in the shared address space, a mapping manager maps the shared address to the physical memory (which can be locate or remote). The mapping manager is a layer of software implemented either in the operating system kernel or as a runtime library routine. To reduce delays due to communication latency, DSM may move data at the shared memory address from a remote node that is accessing data. In such cases DSM makes use of the communication services of the underlying communication.

7.8 Algorithms for Implementing DSM

7.8.1 Implementation issues

a. How to keep track the location of remote data

b. How to minimize communication overhead when accessing remote data

c. How to access concurrently remote data at several nodes

7.8.2 Implementation algorithms

The algorithms for implementing DSM are:

1. The central server algorithm (Fig. 7.6):

   a. Central server maintains all shared data

      i. Read request: Returns data item

      ii. Write request: Updates data and returns acknowledgement message

   b. Implementation

      i. A timeout is used to resend a request if acknowledgment fails

      ii. Associated sequence numbers can be used to detect duplicate write requests

      iii. If an application's request to access shared data fails repeatedly, a failure condition is sent to the application

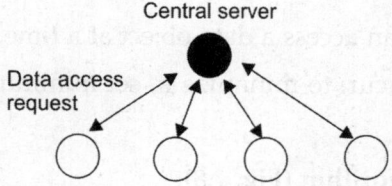

Fig. 7.6: Central server algorithm (schematic)

   c. Issues: Performance and reliability

   d. Possible solutions

      i. Partition shared data between several servers

      ii. Use a mapping function to distribute/locate data

2. The migration algorithm (Fig. 7.7):

a. Operation

i. Ship (migrate) entire data object (page, block) containing data item to requesting location

ii. Allow only one node to access a shared data at a time

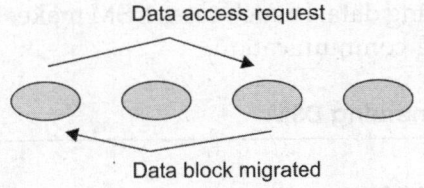

Data access request

Data block migrated

Fig. 7.7: Migration algorithm (schematic)

b. Advantages

i. Takes advantage of the locality of reference

ii. DSM can be integrated with VM at each node

- Make DSM page multiple of VM page size
- A locally held shared memory can be mapped into the VM page address space
- If page not local, fault-handler migrates page and removes it from address space at remote node

c. To locate a remote data object:

i. Use a location server

ii. Maintain hints at each node

iii. Broadcast query

d. Issues

i. Only one node can access a data object at a time

ii. Thrashing can occur: to minimize it, set minimum time data object resides at a node

3. The read-replication algorithm (Fig. 7.8):

a. Replicates data objects to multiple nodes

b. DSM keeps track of location data objects

c. Multiple nodes can have read access or one node write access (multiple readers—one writer protocol)

d. After a write, all copies are invalidated or updated

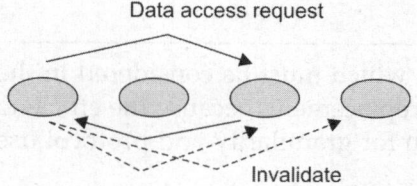

Fig. 7.8: Read-replication algorithm

e. DSM has to keep track of locations of all copies of data objects. Examples of implementation:

   i. IVY: Owner node of data object knows all nodes that have copies

   ii. PLUS: Distributed linked-list tracks all nodes that have copies

f. Advantage

   i. The read-replication can lead to substantial performance improvements if the ratio of reads to writes is large

4. The Full-Replication Algorithm (Fig. 7.9):

a. Extension of read-replication algorithm: Multiple nodes can read and multiple nodes can write (multiple-readers, multiple-writers protocol)

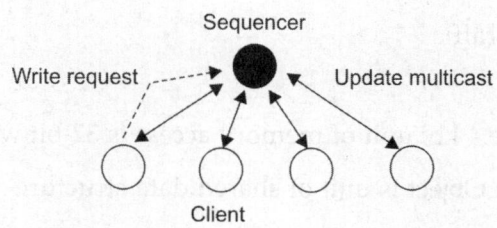

Fig. 7.9: Write operation in full replication algorithm

b. Issue: Consistency of data for multiple writers

c. Solution: Use of gap-free sequencer

   i. All writes sent to sequencer

   ii. Sequencer assigns sequence number and sends write request to all sites that have copies

   iii. Each node performs writes according to sequence numbers

   iv. A gap in sequence numbers indicates a missing write request: node asks for retransmission of missing write requests.

7.9 Design Issues in DSM

The two important issues which must be considered in the design of a DSM system are granularity and page replacement, because the efficiency of DSM depends on the effectiveness of file chosen for granularity and protocol used for page replacement.

7.9.1 Granularity

It refers to the block size of a DSM system, that is to the unit of sharing and the unit of data transfer across the network when a network fault occurs. Hence, can be explained as:

a. If DSM page size is a multiple of the local virtual memory (VM) management page size (supported by hardware), then DSM can be integrated with VM, i.e. use the VM page handling

b. Advantages versus disadvantages of using a large page size:

   i. (+) Exploit locality of reference

   ii. (+) Less overhead in page transport

   iii. (–) More contention for page by many processes

c. Advantages versus disadvantages of using a small page size

   i. (+) Less contention

   ii. (+) Less false sharing (page contains two items, not shared but needed by two processes)

   iii. (–) More page traffic

d. Examples

   i. PLUS: Page size 4 kb, unit of memory access is 32-bit word

   ii. Clouds, Munin: Object is unit of shared data structure

7.9.2 Page replacement

If the local memory of a node is full, a cache miss at that node implies not only a fetch of the accessed data block from a remote node but also a replacement. That is a data block of a local memory must be replaced by new data block. Therefore, a cache replacement strategy is also important in the design of a DSM system. Replacement algorithm (e.g. LRU) must take into account the page access modes: Shared, private, read-only, and writable

Example: LRU with access modes

   i. Private (local) pages to be replaced before shared ones

   ii. Private pages swapped to disk

   iii. Shared pages sent over network to owner

   iv. Read-only pages may be discarded (owners have a copy)

SUMMARY

Distributed file systems are very heavily employed in organizational computing and their performance has been a subject of preferred tuning. NFS has a simple stateless protocol, it has maintained its early position as the dominant distributed file system.

As distributed systems are accepted widely, so both DFS and DSM are important issues as they provide availability, scalability and heterogeneity. There primary goal is to provide physically dispersed users, a common file system that hide the heterogeneity of underlying physical stream. The requirements of DFS are transparency, file replication, etc. while the requirements of DSM are granularity and page replacement.

REVIEW QUESTIONS

1. Explain the implementation algorithms for DSM.
2. Explain the design issues of file system.
3. What do you understand by data integrity?
4. Explain file system with its goal.
5. Write a short notes on architecture of distributed shared memory.
6. Explain:
    a. Mounting
    b. Hints
    c. Caching
7. What are the merits and demerits of distributed shared memory?

UNIT 4

8

Failure Recovery in Distributed System

Contents

8.1 Introduction

- Recovery refers to restoring a system to its normal operational state. Once a failure has occurred, it is essential that the process where the failure happened can recover to a correct state.

- Fundamental to fault tolerance is the recovery from an error.

- Resources are allocated to execute processes in a computer. For example, a process has memory allocated to it and a process may have locked shared resources, such as files and memory.

- Following are some resolutions on process recovery

  - The resources allocated to process

  - Undo modification made to databases, and

  - Restart the process or restart process from point of failure and resume execution

- In distributed process recovery, undo effect of interactions of failed process with other cooperating process.

8.2 General Concepts

- A system is a combination of hardware and software components. These components provide a specified service.
- Failure of a system occurs when the system does not perform its service in the manner specified.
- An erroneous state of the system is a state which could lead to a system failure by a sequence of valid state transitions.
- A system is said to "fail" when it cannot meet its promises. A failure brought by the existence of "errors" in the system.
- A system is said to have a failure, if the service it delivers to the users deviates from compliance with the system specification for a specified period of the system.
- Figure 8.1 shows concept of fault and recovery.
- System failure: System does not meet requirements, i.e. does not perform its services as specified.
- Erroneous system state: State which could lead to a system failure by a sequence of valid state transitions.
- Error: The part of the system state which differs from its intended value.
- Fault: Anomalous physical condition, e.g. design errors, multicasting problems, damage, external disturbances.

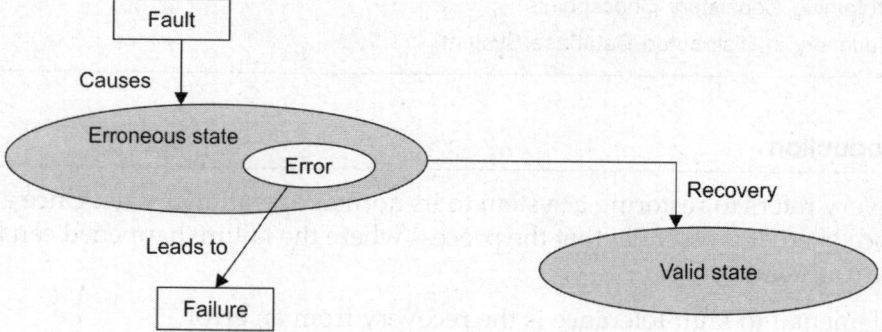

Fig. 8.1: Concept of recovery

8.3 Classification of Failures

Failures in distributed system can be classified as:
1. Process failure
2. System failure
3. Secondary storage failure
4. Communication medium failure

1. **Process failure:** A process proceeds incorrectly (owing to a bug or consistency violation) or not at all (due to deadlock, live lock or an exception). Recovery from process failure involves aborting or restarting the process.

Behaviour: Process causes system state to deviate from specification, e.g. incorrect computation and process stop execution.

Errors causing failure: Protection violation, deadlocks, timeout, wrong user input, etc.

Recovery: Abort process or restart process from the prior state.

2. **System failure:** The processor may fail to execute due to software faults (also known as an OS bug) or hardware faults (affecting the CPU, main memory, the bus, the power supply, a network interface, etc.). Recovery from such failures generally involves stopping and restarting the system. While this may help with intermittent faults, or failures triggered by a specific and rare combination of circumstances, in other cases it may lead to the system failure again at the same point. In such a case, recovery requires external interference (replacing the faulty component) or an automatic reconfiguration of the system to exclude the faulty component (e.g. reconfiguring the distributed computation to continue without a faulty node). It can be further classified as:

   a. **An amnesia failure** occurs when the system restarts in a predefined state that does not depend upon the state of the system before its failure.

   b. **A partial amnesia** failure occurs when the system restarts in a state that wherein a part of the state is same as the state before the failure and rest of the state is predefined.

   c. **A pause failure** occurs when a system restart in the same state it was in before the failure.

   d. **A halting failure** occurs when a crashed system never restarts.

3. **Storage failure:** Some (supposedly) stable storage has become inaccessible (typically a result of hardware failure). Recovery involves rebuilding the device's data from archives, logs, or mirrors (RAID).

   Behavior: Stored data cannot be accessed.

   Error causing failure: Parity error, head crash, etc.

   Recovery/design strategies: Reconstruct content from archive plus log of activities and design mirrored disk system.

4. **Communication medium failure:** Communication fails due to a failure in communications link or intermediate node. This leads to corruption or loss of messages and may lead to partitioning of the network. Recovery from communication medium failure is difficult and sometimes impossible.

   Behavior: A site cannot communicate with another operational site. Errors/faults: Failure of switching node or communication links. Recovery/design strategies: Reroute, error-resistant communication protocols.

8.4 Backward and Forward Error Recovery

Once a failure has occurred, it is essential that the process where the failure happened recovers to a correct state. Recovery is fundamental to fault tolerance. There are two approaches to failure recovery (Fig. 8.2):

1. Forward error recovery:
   - Remove errors in process/system state (if errors can be completely assessed).
   - Continue process/system forward execution.

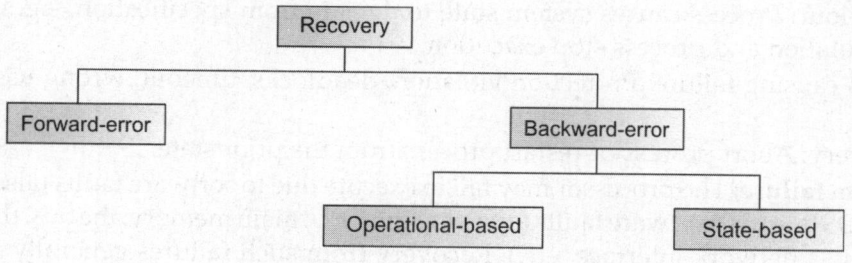

Fig. 8.2: Error recovery

2. Backward error recovery:
   • Restore process/system to previous error-free state and restart from there.

1. **Forward error recovery:** Requires removing (repairing) all errors in the system's state, thus enabling the processes or system to proceed. No actual computation is lost (although the repair process itself may be time consuming). Forward recovery implies the ability to completely assess the nature of all errors and damages resulting from the faults that lead to failure. An example could be the replacement of a broken network cable with a functional one. Here it is known that all communication has been lost, and if appropriate protocols are used (which, for example, buffer all outgoing messages) a forward recovery may be possible (e.g. by resending all buffered messages). In most cases, however, forward recovery is impossible.

   Advantage: i. Less overhead
   Disadvantages: i. Limited use, i.e. only when impact of faults understood
                      ii. Cannot be used as general mechanism for error recovery

2. **Backward error recovery:** This restores the process or system state to a previous state known to be free from errors, from which the system can proceed (and initially retrace its previous steps). Obviously this incurs overheads due to the lost computation and the work required to restore the state. Also, there is in general no guarantee that the same error will not reoccur (e.g. if the failure resulted from a software bug). Furthermore, there may be irrecoverable components, such as external input (from humans) or irrevocable outputs (e.g. cash dispensed from an ATM). While the implementation of backward recovery faces substantial difficulties, it is in practice the only way to go, due to the impossibility of forward recovery from most errors. For the remainder of this lecture we will, therefore only look at backward recovery.

   Advantages: i. Simple to implement
                       ii. Can be used as general recovery mechanism
   Disadvantages: i. Performance penalty
                      ii. No guarantee that fault does not occur again
                     iii. Some components cannot be recovered

8.5 Backward error Recovery

This relies on restoring the system to a previous safe state executing an alternative section of the program. This has the same functionality but uses a different algorithm and therefore no fault, i.e. it works by restoring processes to a recovery point, which represents a pre failure state of the process. A system can be recovered by restoring all its active processes to their recovery points. The recovery can be done at process level or system level.

Process level recovery: It is simply subset of the actions necessary to recover the entire system.

System level recovery: All the user processes that were active need to be restored to their respective recovery points and data modified by the process need to restore at a proper state.

Recovery can happen in one of two ways:

i. **Operation-based recovery** keeps a log (or audit trail) of all state-changing operations. The recovery point is reached from the present state by reversing these operations;

ii. **State-based recovery** stores a complete prior process state (called a checkpoint). The recovery point is reached by restoring the process state from the checkpoint (called rollback). State-based recovery is also frequently called rollback recovery.

System model

Consider a single machine system. Machine consists of secondary storage and stable storage systems. Stable storage means it does not loose information if system fails. Bring object to main memory to be accessed (Fig. 8.3). Then it stores logs and recovery points. If the access is a write operation, the copy of the object in the main memory is updated.

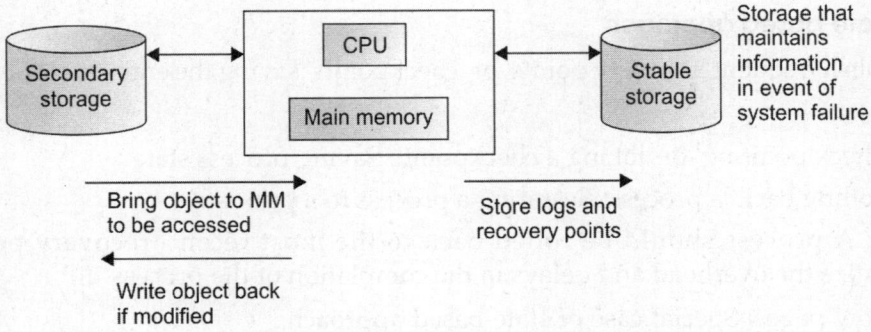

Fig. 8.3: System model

The data object in the secondary storage is updated when the copy of the object in main memory is flushed to the disk by paging scheme. Stable storage is used to store logs and recovery points.

8.5.1 Operation based approach

- Record all changes made to state of process ('audit trail' or 'log') such that process can be returned to a previous state
- Example: A transaction based environment where transactions update a database
  a. It is possible to commit or undo updates on a per transaction basis
  b. A commit indicates that the transaction on the object was successful and changes are permanent

Updating-in-place

- Principle: Every update (write) operation to an object creates a log in stable storage that can be used to *'undo'* and *'redo'* the operation
- Log content: Object name, old object state, new object state
- Implementation of a recoverable update operation:
  1. *Do* operation: Update object and write log record
  2. *Undo* operation: Log (old) → object (undoes the action performed by a *do*)
  3. *Redo* operation: Log (new) → object (redoes the action performed by a *do*)
  4. *Display* operation: Display log record (optional)
- Problem: A *'do'* cannot be recovered if system crashes after write object but before log record write.

The write-ahead log protocol

- Principle: Write log record before updating object.
- It is implemented by two operations:
  a. Update the object only after the undo log is recorded.
  b. Before committing the updates, redo and undo logs are recorded.

8.5.2 State based approach

- Establish frequent 'recovery points' or 'checkpoints' saving the entire state of process
- Actions:
  1. 'Check pointing' or 'taking a checkpoint': Saving process state
  2. 'Rolling back' a process: Restoring a process to a prior state

  Note: A process should be rolled back to the most recent 'recovery point' to minimize the overhead and delays in the completion of the process
- Shadow pages: Special case of state-based approach
  1. Only a part of the system state is saved to minimize recovery
  2. When an object is modified, page containing object is first copied on stable storage (shadow page)
  3. If process successfully commits: Shadow page discarded and modified page is made part of the database
  4. If process fails: Shadow page used and the modified page discarded.

8.6 Recovery in Concurrent Systems

- The term recovery refers to the process of restoring a (failed) system to a normal state of operation. Recovery can apply to the complete system (involving rebooting a failed computer) or to a particular application (involving restarting of a failed process(es)).

- If one of a set of cooperating processes fails and has to be rolled back to a recovery point, all processes it communicated with since the recovery point have to be rolled back.

- Information exchange can be through a shared memory in the case of shared memory machines.

- In concurrent and/or distributed systems all cooperating processes have to establish recovery points

8.6.1 Orphan messages and the domino effect

The process of a cascaded rollback may lead to what is called the *domino effect* (Fig. 8.4). An *orphan message* is a process that survives the crash of another process, but whose state is inconsistent with the crashed process after its recovery. Three processes X, Y and Z exchange their information. Information is shown by $\rightarrow$ and symbol "[" marks a recovery point to which a process can be rolled back in the event of a failure.

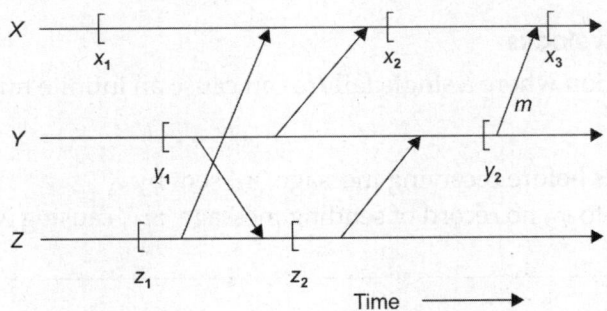

Fig. 8.4: Domino effect

Case 1: Failure of X after x_3 : No impact on Y or Z

Case 2: Failure of Y after sending 'm'
- Y rolled back to y_2
- 'm' a orphan massage
- X rolled back to x_2

Case 3: Failure of Z after z_2
- Y has to rollback to y_1
- X has to rollback to x_1 (domino effect)
- Z has to rollback to z_1

8.6.2 Lost messages

Regenerating lost messages (Fig. 8.5) on recovery:

1. If implemented on unreliable communication channels, the application is responsible.
2. If implemented on reliable communication channels, the recovery algorithm is responsible

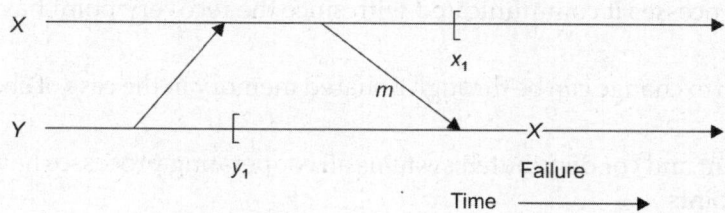

Fig. 8.5: Lost messages

- Assume that x_1 and y_1 are the only recovery points for processes X and Y, respectively
- Assume Y fails after receiving message 'm'
- Y rolled back to y_1, X rolled back to x_1
- Message 'm' is lost
- There is no distinction between this case and the case where message 'm' is lost in communication channel and processes X and Y are in states x_1 and y_1, respectively.

8.6.3 Problem of livelocks

Livelock is a situation where a single failure can cause an infinite number of rollbacks.

Case 1 (Fig. 8.6)

   i. Process Y fails before receiving message 'n_1' sent by X
   ii. Y rolled back to y_1, no record of sending message 'm_1', causing X to rollback to x_1.

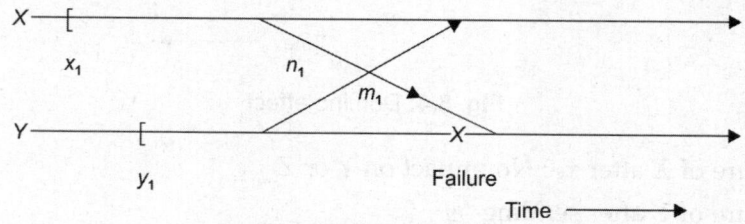

Fig. 8.6: Case 1—livelock

Case 2 (Fig. 8.7)

- When Y restarts, sends out 'm_2' and receives 'n_1' (delayed).
- When X restarts from x_1, sends out 'n_2' and receives 'm_2'.
- Y has to rollback again, since there is no record of 'n_1' being sent.
- This cause X to be rolled back again, since it has received 'm_2' and there is no record of sending 'm_2' in Y.
- The above sequence can repeat indefinitely.

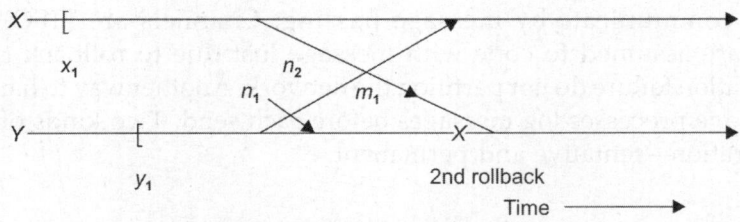

Fig. 8.7: Case 2—livelock

8.7 Obtaining Consistent Checkpoints

- Checkpointing in distributed systems require all processes (sites) that interact with one another to establish periodic checkpoints
- All the sites save their local states: *Local checkpoints*
- All the local checkpoints, one from each site, collectively form a *global checkpoint*
- The domino effect is caused by orphan messages, which in turn, are caused by rollbacks.

Strongly consistent set of checkpoints

- The most recent distributed snapshot in a system is also called the recovery line. It corresponds to a strongly consistent global state.
- Establish a set of local checkpoints (one for each process in the set) such that no information flow takes place (i.e. no orphan messages) during the interval spanned by the checkpoints.
- There is one recovery point for each process in the set during the interval spanned by checkpoints; there is no information flow between any pair of processes in the set and any process outside the set.

Consistent set of checkpoints

- A process saves its local state on the stable storage, which is called a local checkpoint.
- The process of saving local states is called local checkpointing.
- All the local checkpoints, one from each site, collectively form a global checkpoint.
- A global checkpoint is a strongly consistent set of checkpoints if there is no orphan message and no lost message.
- A global checkpoint is a consistent set of checkpoints if there is no orphan message
- Each message that is received in a checkpoint (state) should also be recorded as sent in another checkpoint (state).

8.7.1 Synchronous checkpointing

- Livelock problem during recovery is avoided by taking a consistent set of checkpoints. Algorithm is said to be synchronous when the processes involved coordinate their local checkpointing actions such that the set of all recent checkpoints in the system guaranteed to be consistent.

- Processes communicate by message passing. Channels are FIFO. End-to-end protocols are assumed to cope with message lost due to rollback recovery and communication failure do not partition the network. Another way to handle message loss is to have processes log messages before each send. Two kinds of checkpoints are in operation—tentative and permanent.

First phase

- The initiator requests all processes to take tentative checkpoints.
- Each process informs the initiator whether it would like to take a tentative checkpoint. A process will not say "yes" if there is a failure or other reasons.

Second phase

The initiator requests them to make tentative checkpoints permanent if it receives "yes" from all of them; otherwise asks them to discard the tentative checkpoints.

Optimization

- Sometimes, it is not necessary to ask all processes to take checkpoints for each checkpointing initiation.
- Each message is attached to a monotonically increasing number.
- Let m be the last message that X received from Y after X has taken its last checkpoint, Then, $last\_rcvdx[y] = m.l$, if m exists, otherwise, it is $\wedge$.
- Let m be the first message that X sent to Y after X took its last checkpoint. Then, $first\_sentx[y] = m.l$, if m exists, otherwise, it is $\wedge$.
- Whenever X request Y to take a tentative checkpoint, X sends $last\_rcvdx[y]$ along with its request; Y takes a checkpoint only if $last\_rcvdx[y]^3 first\_senty[x] > \wedge$ cohortx= $\{y \mid last\_rcvdx[y] > \wedge\}$
- Initial state at all processes p; – for all processes q do $first\_sentp[q] = \wedge$; – OK_cpp="yes", if p is willing to take a checkpoint, "no" otherwise.

8.7.2 Asynchronous checkpointing

- Under asynchronous approach, checkpoints at each process are taken independently without synchronization.
  - Remove the synchronization overhead of the coordinated approach.
  - May result in domino effect.
- One solution is to log incoming messages.
  - Pessimistic message logging: An incoming message is logged before it is processed. Too much overhead slows down computation.
  - Optimistic: Processes continue to perform the computation, and store messages in memory, which will be logged at certain intervals. More rollbacks during failure may still have domino effects.

- The communication channels are reliable, FIFO, and have infinite buffers. The message communication delay is arbitrary, but finite.

- The underlying computation is assumed to be event-driven, where a process waits until a message is received, processes the message, changes its state, and sends messages to other processes.

8.8 Recovery in Distributed Database System

To enhance performance and availability, a distributed database system is replicated where copies of data objects are stored at different sites. Such system is known as a replicated distributed database system.

Replication: System maintains multiple copies of data, stored in different sites, for faster retrieval and fault tolerance. A relation or fragment of a relation is replicated if it is stored redundantly in two or more sites. Full replication of a relation in the case where the relation is stored at all sites. Fully redundant databases are those in which every site contains a copy of the entire database. The availability and performance of a database system is enhanced as the transactions are not blocked even when one or more site fails. Goal of recovery algorithm in RDDBS is to hide inconsistencies from user transaction, bring the copies at recovering sites up-to-date with respect to the rest of the copies.

- **Advantages of replication**

  1. **Availability:** Failure of site containing relation r does not result in unavailability of r as replicas exist.

  2. **Parallelism:** Queries on the system may be processed by several nodes in parallel.

  3. **Reduced data transfer:** Relation r is available locally at each site containing a replica of r.

- **Disadvantages of replication**

  1. **Increased cost of updates**: Each replica of relation r must be updated.

  2. **Increased complexity of concurrency control:** Concurrent updates to distinct replicas may lead to inconsistent data unless special concurrency control mechanisms are implemented.

- **Two methods are used to recover the failed sites:**

  1. Message spoolers are used to save all the updates toward failed sites. On recovery, the failed site processes all the missed updates before resuming normal operations.

  2. Second method uses special transactions known as copier transaction. Copier transaction reads the up-to-date copies at the operational site and updates the copies at recovering sites.

SUMMARY

The term recovery refers to the process of restoring a (failed) system to a normal state of operation. Recovery can apply to the complete system (involving rebooting a failed computer) or to a particular application (involving restarting of failed processes. While restarting processes or computers are a relatively straightforward exercise in a centralized system, things are (as usual) significantly more complicated in a distributed system. The main challenge for implementing recovery in a distributed system arises from the requirement of consistency. An isolated process can easily be recovered by restoring it to a checkpoint. A distributed computation, however, consists of several processes running on different nodes. If one of them fails, it may have causally affected another process running on another node. A process B is causally affected by process A if any of A's state has been revealed to B (and B may have subsequently based its computation on this knowledge of A's state). Recovery can be obtained by backward and forward recovery approach, using the concept of checkpointing.

REVIEW QUESTIONS

1. Explain the fault and error.
2. Explain the classification of failures.
3. List the advantages and disadvantages of forward and backward recovery.
4. Explain synchronous checkpointing and recovery.
5. Write a short note on recovery in distributed database system.
6. Explain the concept of recovery in concurrent system.
7. Discuss the two approaches of backward recovery.

9

Fault Tolerance

Contents

9.1 Introduction

Fault tolerance is defined as the ability of the system to avoid disruption due to failures and to improve availability. The purpose of fault tolerance is to increase the dependability of a system. A complementary but separate approach to increasing dependability is fault prevention. This consist of techniques such as inspection whose intent is to eliminate the circumstances by which fault arise. Fault tolerance is a non-functional (QoS) requirement that requires a system to continue operation, even in the presence of faults, fault tolerance should be achieved with minimal involvement of users or system administrators. Distributed systems can be more fault tolerant than centralized systems, but with more processor hosts generally the occurrence of individual faults is likely to be more frequent.

There are two approaches to fault tolerance:

1. **Mask failures:** System continues to provide its specified functions in the presence of failures.

2. **Well-defined failure behavior:** System may not perform its specified functions but behaves/exhibits a well-defined behavior in presence of failures.

9.2 Issues in Fault Tolerance

9.2.1 Process deaths

In this case both clients and servers should be informed in case of failures so that they can be unblocked and the system may not fail and also the resources allocated to that process are recouped, else they may be lost permanently.

9.2.2 Machine Failures

In this, the process running at the machine will die although the behavior of client process or a server process does not show much difference in the event of machine failure.

9.2.3 Network failure

In this, the communication link failure occurs which partitions the network into subnets and hence our fault tolerance system should be designed such that it may assume that a machine may be operating and processes on that machine are active.

The faults in distributed system can be differently classified on the basis of attributes, i.e. they can be classified by their underlying cause. These are faults that occur during lifetime of the system and are invariably due to physical causes such as processor failure or disk crashes, are described in Table 9.1.

Table 9.1: Types of failure	
Type of failure	*Description*
Crash failure	A server halts, but is working correctly until the failure
Amnesia crash	Lost all history, must reboot
Pause crash	Still remember state before crash, can be recovered
Halting crash	Hardware failure, must be replaced or re-installed
Omission failure	A server fails to respond to incoming requests
Receive omission	A server fails to receive incoming messages
Send omission	A server fails to send messages
Timing failure	A server's response lies outside the specified time interval
Response failure	The server's response is incorrect
Value failure	The value of the response is wrong
State transition failure	The server deviates from the correct flow of control
Arbitrary failure	A server may produce arbitrary responses at arbitrary times

9. 3 Commit Protocols

Commit protocols are used to ensure atomicity across sites:
- A transaction which executes at multiple sites must either be committed at all the sites, or aborted at all the sites.
- Not acceptable to have a transaction committed at one site and aborted at another.

General Paradox: There are two generals of the same army who have encamped a short distance apart. Their objective is to capture a hill, which is possible only if they attack simultaneously. If only one general attacks, he will be defeated. The two generals can only communicate by sending messengers, which is not reliable. The challenge is to use a protocol that allows the generals to agree on a time to attack, even though some messengers do not get through.

Assume there is a protocol which sends messengers a fixed number of times to solve the problem. Let P be the shortest protocol. Suppose the last messenger in P does not reach the destination. Then either the message carried by the messenger is useless or one of the generals does not get the needed message. Since P is the minimum length protocol by our assumption, the message that was lost was not a useless message and hence one of the generals will not attack.

9.3.1 Two phase commit protocols

It is a distributed algorithm which allows all the sites to agree whether to commit a process or to abort (in case of site failure) a transaction. The two phases of the algorithm are (Fig. 9.1):

1. The COMMIT_REQUEST phase, where the coordinator attempts to prepare all the cohorts, and
2. The COMMIT phase, where the coordinator completes the transactions at all cohorts.

Phase 1

At the coordinator:

1. The coordinator sends a COMMIT_REQUEST to every cohort requesting the cohorts to commit
2. The coordinator waits for replies from all the cohorts.

At cohorts:

1. On receiving the COMMIT_REQUEST, if the transaction executing is successful, it writes UNDO and REDO log and sends an agree to the coordinator, otherwise, it sends an ABORT.

Phase 2

At the coordinator:

1. If all cohorts agree, the coordinator writes a COMMIT into the log. Then, it sends a commit to all cohorts. Otherwise, it sends an abort.
2. The coordinator waits for acknowledgement from each cohort.
3. If an acknowledgement is not received from any cohort within a timeout period, it resends the COMMIT / ABORT to that cohort.
4. If all the acknowledgements are received, it writes a complete to the log.

At cohorts:

1. On receiving a commit, it releases locks for executing the transaction, and sends an acknowledgement.
2. On receiving an abort, it undoes the transaction using UNDO, releases the locks and sends an ACK message.

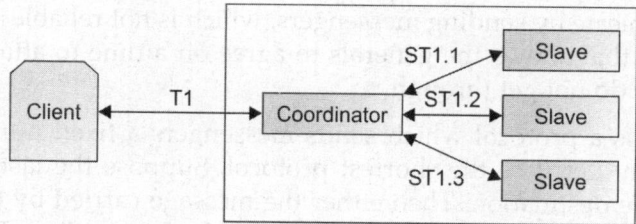

Fig. 9.1: Two-phase commit protocol architecture

Site Failures: In case of message loss, resend.

- Coordinator crashes before writing COMMIT
  - On recovery, it broadcasts an ABORT, cohorts who had agreed to commit, UNDO the transaction and abort. Other cohorts simply undo the transaction. Note that cohorts are blocked until they receive an ABORT.
- Coordinator crashes after writing commit but before writing complete
  - On recovery, it broadcasts a COMMIT and waits for acknowledgement. Cohorts are blocked until they receive a COMMIT.
- Coordinator rashes after writing complete.
  - Nothing to be done on recovery.
- If a cohort crashes in phase 1, the coordinator will abort the transaction.
- Suppose a cohort crashes in phase II, i.e., after writing UNDO and REDO.
  - On recovery, the cohort will check with the coordinator whether to abort/commit.
  - Committing may require a REDO operation because the cohort may have failed before updating the database.

9.3.2 Non-blocking commit protocols

If transactions resilient to site failures, the commit protocols must not block, in the event of site failures. To ensure that commit protocols are non-blocking in the event of site failure, operational sites should agree on the outcome of the transaction by examining their local sites. In addition, the failed sites, upon recovery must reach the same conclusion regarding the outcome of the transaction. This decision must be consistent with the final outcome at the sites that are operational. This protocol is synchronous within one state transition if one site never leads another site by more than one state transition (Fig. 9.2).

Assumptions: The communication network is assumed to have the following characteristics:

  i. The network is reliable and point to point communication is possible between any 2 operational sites.
  ii. The network can detect the failure of a site and report it to the site trying to communicate with the failed site.

Phase 1

1. The coordinator sends VOTE_REQ to all participants.
2. When a participant receives VOTE_REQ, it responds with YES or NO, depending on its vote. If a participant votes NO, it decides abort and stops.

Phase 2

1. The coordinator collects all votes. If any vote was NO, it decides abort and sends ABORT to all the participants that voted YES, and stops. Otherwise, it sends PRE_COMMIT message to all the participants.
2. A participant that voted YES waits for a PRE_COMMIT or ABORT message from the coordinator. If it receives a PRE_COMMIT, then it responds with an ACK message.

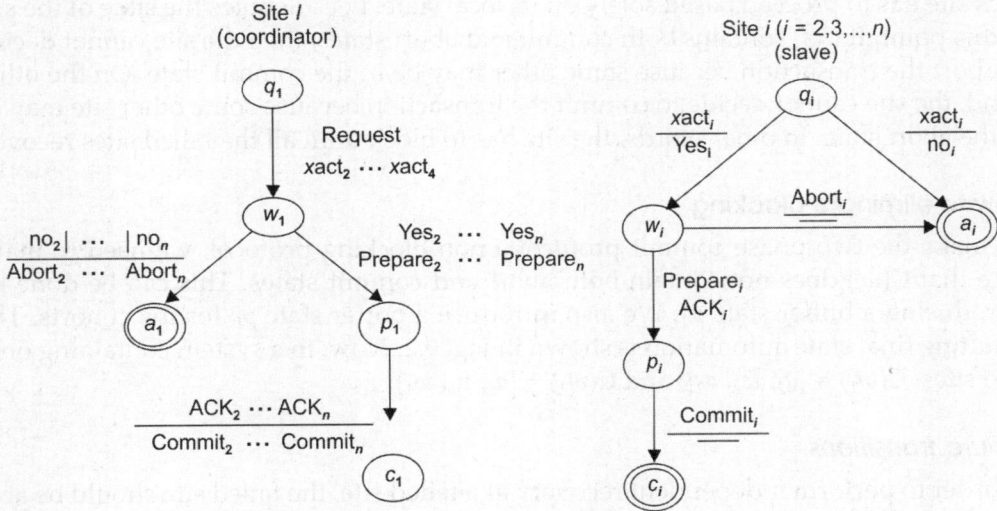

Fig. 9.2: Finite state diagram of three-phase commit protocol

Phase 3

1. The coordinator collects the ACKs. When they have all been received, it decides commit, sends COMMITs, to all participants, and stops.
2. A participant waits for a COMMIT from the coordinator. When it receives that message, it decides commit and stops.

Definitions

Synchronous protocols

A protocol is said to be *synchronous* within one state transition if one site never leads another site by more than one state transition during the execution of the protocol.

Concurrency set

Let s_i denotes the state of a site$i$. The set of all the states of every site that may be concurrent with it is known as the concurrency set of si (denoted by $C(s_i)$. For example, consider a system having two sites, using the two-phase commit protocol. If site 2's state is w_2, then $C(w_2) = \{c_1, a_1, w_1\}$. Likewise, $C(q_2) = \{q_1, w_1\}$. Note that, a_1, c_1 do not belong to $C(q_2)$ because the two-phase commit protocol is synchronous within one state transition only.

Sender set

Let s be an arbitrary state of a site, and let M be the set of all messages that are received in state s. The sender set for s, denoted $S(s) = \{i \mid \text{site } i \text{ sends } m \text{ and } m \text{ belongs to } M\}$

Conditions that cause blocking

Let us see the conditions that lead to the two-phase commit protocol blocks. Consider a simple case where only one site remains operational and all other sites have failed. This site has to proceed based solely on its local state. Let s denotes the state of the site at this point. If $C(s)$ contains both commit and abort states, then the site cannot decide to abort the transaction because some other may be in the commit state. On the other hand, the site cannot decide to commit the transaction because some other site may be in the abort state. In other words, the site has to block until all the failed sites recover.

How to eliminate blocking

To make the two-phase commit protocol a non-blocking protocol, we need to make sure that $C(w_i)$ does not contain both abort and commit states. This can be done by introducing a buffer state p_1. We also introduce a buffer state p_i, for the cohorts. The resulting final state automation is shown in Fig. 9.2. Now, in a system containing only two sites, $C(w_1) = \{q_2, w_2, a_2\}$, and $C(w_2) = \{a_1, p_1, w_1\}$.

Failure transitions

In order to perform independent recovery at a failed site, the failed site should be able to reach a final decision based solely on its local state. A *failure transition* occurs at a failed site at the instant it fails (or immediately after it recovers from the failure). The site upon recovery assumes this local state caused by the failure transition. The transitions are performed according to the following rules:

Rule 1

For every non-final state s (i.e. q_i, w_i, p_i) in the protocol, if $C(s)$ contains a commit, then assign a failure transition from s to a commit state in its FSA; otherwise, assign a failure transition from s to an abort state in its FSA.

Timeout transitions

Operational sites perform a timeout transition in the event of another site's failure. If site i is waiting for a message from site j (i.e. j belongs to $S(i)$) and site j has failed, then site i times out. Based on the type of message from j, we can determine in what state site j failed. Once the state of j in known, we can determine the final state of j due to the failure transition at j. This observation leads to the timeout transitions in the commit protocol at the operational sites.

Rule 2

For each non-final state s, if site j is in $S(s)$, and site j has a failure transition to a commit (abort) state, then assign a timeout transition from state s to a commit (abort) state in the FSA.

The rationale behind this rule is as follows. The failed site makes a transition to a commit (abort) state using the failure transition (Rule 1). Therefore, operational sites must make the same transition in order to ensure that the final outcome of the transaction is identical at all the sites.

9.4 Voting Protocols

Principles

1. Data replicated at several sites to increase reliability.

2. Each replica assigned a number of votes.

3. To access a replica, process must collect a majority of votes.

   The voting mechanism is more fault tolerant than a commit protocol, in that it allow access to data network partitions, site failures and message losses. The voting mechanism can be of two types:

   1. Static voting

   2. Dynamic voting

9.4.1 Static voting

It was proposed by Gifford.

System model

1. File replicas at different sites. The file lock rule: One writer and no reader or multiple readers and no writer.

2. Every file is associated with a version number that gives the number of time a file has been updated. Every successful write updates version number.

Basic idea

The essence of a voting algorithm which controls access to replicated data. Every replica is assigned a certain number of votes. This information is stored on stable storage. A read or write operation to permit if a certain number of votes, read quorum or write quorum respectively are collected by the requesting process.

Algorithm

- Site i wishes to read/write x

  1. Requests a lock from its local lock manager.

  2. Sends vote request to all other sites.

  3. Collects (votes, version #) from all the replying sites.

  4. Determines if it has a read/write quorum.

  5. Ensures that its local copy is current.

6. Performs all its local operations on X and then updates, if it got a write.

7. Quorum, all (current) replicas that replied.

8. Releases its local lock, and

9. Sends release-lock messages to all the nodes that replied.

9.4.2 Dynamic voting

The network partitions and other failures may hurt fault tolerance of static voting and also the number of votes or the set of sites that form a quorum changes with the state of system. Dynamic voting can boost fault tolerance by adapting:

a. The number of votes assigned to various nodes.

b. The set of nodes that can form read/write operation.

The two approaches which are used to enhance availability of dynamic voting protocol are:

1. Majority based approach: The set of sites that can form a majority to allow access to replicated data changes dynamically

2. Dynamic vote reassignment: The number of votes assigned to a site changes dynamically.

1. Majority-based approach

Set of sites that can form a majority to allow access to replicated data of changes with the changing state of system. It modifies the read/write quorums so that site in a distinguished network, partition can still operate. A network partition can still operate. A network partition is distinguished if it contains a majority of the nodes that participated in the last update operation (Fig. 9.3).

Example:

- Consider a system with six sites A, B, C, D, E, and F, each with one vote. Let $w = 4$, $r = 3$.

- If only site B fails, operations can still be performed.

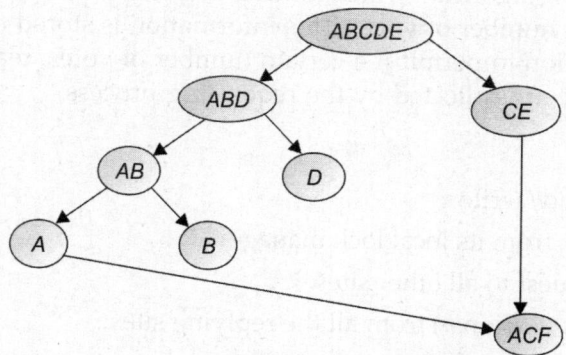

Fig. 9.3: Example of majority-based approach

- If the network is divided into two groups {*ABCD*} and {*EF*}. Both read and write can be performed in the first group, but no operation in the second group.

- If the network is divided into {*ABC*} and {*DEF*}, only read can be performed.

- If the network is divided into {*AB*}, {*CD*}, and {*EF*}, no operation can be performed.

Notations used

- **Version number:** denoted by VNi counts the number of updates to site i. Initialized to 0.

- **Number of replicas updated**. Denoted by RUi, which is the number of replicas participating in the most recent update. Initialized to the number of replicas.

- **Distinguished site list**, denoted by DSi, which stores IDs of one or more sites.
  - When RUi is even, DSi identifies the replica with the largest id number.
  - When RUi is odd, DSi is empty except when $RUi = 3$, in which case DSi lists the three replicas.

Outline of protocol

1. Site i issues a lock_request to its local lock manager. If the lock is granted, i will send a vote_request to all the sites.

2. When a site j receives the vote_request, it obtains locks and sends the values of VNj, RUj, and DSj to site i.

3. From the responses, site i decides whether it belongs to the distinguished partition, (described later).

4. If i does not belong to the distinguished partition, it releases locks and sends abort to the responded sites, who also releases locks.

5. If i belongs to the distinguished partition, it gets the current data copy and update. Also i execute the update protocol. Site i then send a commit to all participating sites, and asks them to update the data, VNj, RUj, and DSj. It then releases the locks.

6. When a site j receives the commit message, it updates the data, and VNj, RUj, and DSj, then releases the locks.

2. Dynamic vote reassignment protocol

This protocol is divided into two types:

a. Group Consensus: The sites in the active group agree upon the new vote assignment using either a distributed algorithm or by selecting a coordinator to perform the task. Since outside the majority group did not receive any votes. Because this method relies on active group participation, the current system topology will be known before deciding the vote assignments.

b. Autonomous Reassignment: Each node makes its own decision about changing its votes and picking a new vote value, without regarding rest of nodes. Before the change is made final, though, node must collect majority of votes.

SUMMARY

In distributed systems, it is necessary to have a mutual exclusion mechanism that works even when does tail or the communication lines are broken. For example, consider a system that manages replicated data. Owing to a network partition, the system may be divided into isolated group of nodes, we probably do not want users at isolated groups updating the database concurrently since this would cause the copies to diverge. So, if a group is going to perform updates, it must be able to guarantee that no other group is performing this activity. This mutual exclusion has to be enforced without communication between groups. In our recent discussion, we studied how different commit protocol variants, enhanced with optimizations or not, can coexist in the form of a universal commit protocol that allows the execution of distributed transactions in heterogeneous systems. This universal commit protocol is called "Al/two-phase commit" (ALL2PC). This work resulted in 3 BS theses (the last publication below) at the University of Cyprus and is in preparation for conference and journal publication. In distributed database and transaction systems a distributed commit protocol is required to ensure that the effects of a distributed transaction are atomic, that is, either all the effects of the transaction persist or none persist, whether or not failures occur. Several commit protocols have been proposed in the literature. These are variations of what has become a standard and known as the two-phase commit (2PC) protocol. In active replication model, there are multiple (group) replica managers (RM), each with equivalent roles. The RM's operate as a group and each front end (client interface) multicasts requests to a group of RM's. Requests are processed by all RM's independently (and identically). Client interface compares all replies received and can tolerate N out of 2N + 1 failure, i.e. consensus when N + 1 identical response received. This model also can tolerate Byzantine failure.

REVIEW QUESTIONS

1. What do you understand by fault tolerance in distributed system?
2. What are the different states of fault classification?
3. Write down the various issues in fault tolerance.
4. How to assign votes in distributed system?
5. Explain two-phase commit protocol.
6. Write short notes on:
    a. Passive replication
    b. Non-blocking protocol
    c. Static voting

UNIT 5

10

Transaction and Concurrency Control

Contents

10.1 Introduction

Transaction is a logical unit work on a database which provides the entire series of steps which are necessary to accomplish the task (i.e. logical unit of work). It must see a consistence database. So, transaction is a sequence of reads and writes on the consistent database.

As we know that concurrent execution of user program is essential for good DBMS performance. Because disk access is frequent, and relatively slow, it is important to keep the CPU humming by working on several user programs concurrently. Hence transaction is the DBMS abstract view of a user program, i.e. it can be well defined as the sequence of reads and writes.

A transaction is the execution of a sequence of actions on a server that must be either entirely completed or aborted, independently of other transactions. The main goal of transaction is to provide the following:

i. To allow the execution of concurrent transactions, yet to maintain consistency.

ii. To deal with the failure of either the server or the client.

Examples of transaction:

i. Updating the record.

ii. Bring record into buffer

iii. Locate the record into disk.

iv. Update data into buffer.

v. Writing data back on disk.

121

A simple example of transaction (Fig. 10.1) specifying a series of related actions involving bank accounts A, B and C.

Transaction T

a. Withdraw (500);

b. Submit (500);

c. Withdraw (200);

d. Submit (200);

Fig. 10.1: A transaction

10.2 Basic Concepts

10.2.1 ACID properties of a transaction

Atomic: All parts of transaction must be completed and committed or it must be aborted and rolled back.

Consistent: Each user is responsible to ensure that their transaction would leave the database in a consistent state.

Isolation: The final effects of multiple simultaneous transactions must be the same as if they were executed one right after the other.

Durability: If a transaction has been committed, the DBMS must ensure that its effects are permanently recorded in database.

10.2.2. Transaction state

1. Active, the initial state, the transaction stays in this state while it is executing.
2. Partially committed, after the final statement has been executed.
3. Failed, after the discovery that normal execution can no longer proceed.
4. Aborted, after the transaction has been rolled back and the database restored to its state prior to the start of transaction. Restart the transaction, only if no internal logical error or kill the transaction.
5. Committed, after successful completion (Fig. 10.2).

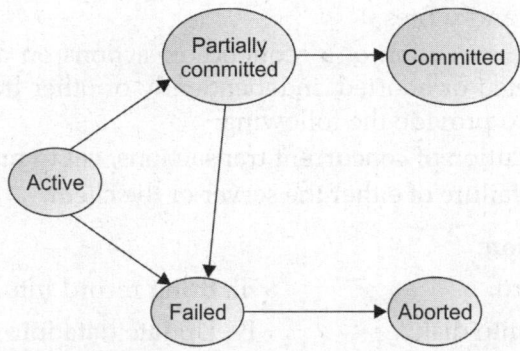

Fig. 10.2: State diagram of a transaction

10.2.3 Transaction log

• Keeps track of all transactions that update the database. Used for recovery in case of a rollback.

1. Record for the beginning of the transaction.

2. Types of operation (insert/update/delete).

3. Names of objects/tables affected by the transaction.

4. Before and after values for updated fields.

5. Pointers to previous and next transaction log entries for the same transaction.

6. The ending of the transaction (commit).

• Used for recovery in case of a rollback.

10.2.4 Flat transaction

It represents the simplest type of transaction, and for almost all existing systems, it is the only one which is supported at the application programming level. It forms the basic building block for organizing an application into atomic actions. It strictly satisfies the ACID properties (Fig. 10.3).

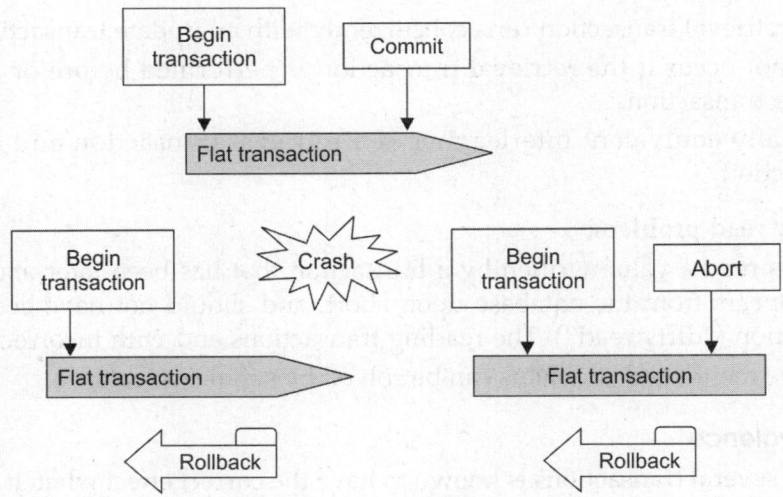

Fig. 10.3: Flat transaction

Limitations of flat transactions

a. Do not model real life workflow and dependencies well.

b. Do not integrate well into mixed models with other actors (other computers outside of transactional domain or even humans).

c. Trip planning: Commit part of a transaction.

d. Bulk updates: Can be expensive to undo all updates.

10.3 Concurrency Control

It is required to manage concurrent, i.e. simultaneous transaction execution in a multiprocessing database system and it also ensures serializability in a multi-user database. The lack of concurrency control may create data integrity and consistency problems like:

1. Lost updated
2. Inconsistent retrievals
3. Dirty reads

It illustrates two well-known problems of concurrent transaction in the context of banking. These can be avoided using serially equivalent execution of transactions.

1. Lost updates:

This problem occurs when:

   i. Two transactions reads the old value of a variable.

   ii. Then use it to calculate the new value.

   iii. This cannot happen if one transaction is performed before the other, because the later transactions will read the value written by the earlier one.

2. Inconsistent retrievals:

The problem occurs when:

   i. If the retrieval transaction runs concurrently with an update transaction.

   ii. It cannot occur if the retrieval transaction is performed before or after, then update transaction.

   iii. A serially equivalent interleaving of a retrieval transaction and an update transaction.

3. The dirty read problem:

Transactions read a value written by a transaction that has been later aborted. This value disappears from the database upon abort, and should not have been read by any transaction ("dirty read"). The reading transactions end with incorrect results.

The above mentioned problems can be solved by serial equivalence.

Serial Equivalence

1. If each of several transactions is known to have the correct effect when it is done on its own, then we can infer that if these transactions are done one at a time in some order, the combined effect will also be correct.

2. An interleaving of the operations of transaction, in which the combined effect is the same as if the transactions had been performed one at a time in some order is a serially equivalent interleaving.

10.4 Nested Transactions

It extends the transaction model by allowing transaction to be composed of other transactions. The outermost transaction in the set is known as top-level transaction and all other transactions are known as sub-transactions (Fig. 10.4).

Hierarchy: The outermost transaction in a set of nested transaction is called the top-level transaction. The transactions other than top-level transactions are known as sub-transactions.

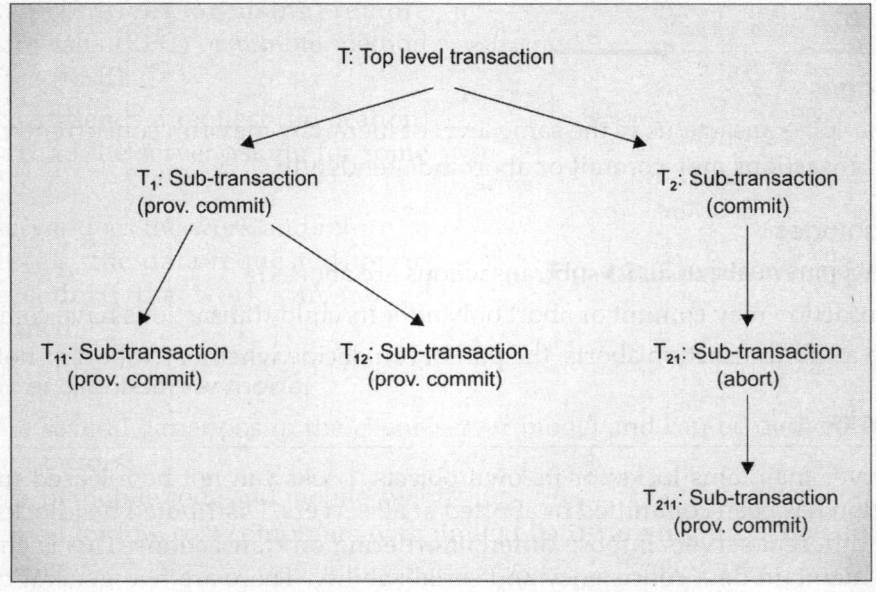

Fig. 10.4: Nested transactions

- T is the top-level transaction which starts a pair of sub-transactions T_1 and T_2.
- The sub-transactoin T_1 starts its own pair of sub-transactions T_{11} and T_{12}.
- The sub-transactoin T_2 starts its own pair of sub-transactions T_{21}, which again starts another sub-transactoin T_{211}.

Properties for nested transactions are:

1. A sub-transaction appears atomic to its parent with respect to the transaction failures and to concurrent access.
2. Sub-transactions at the same level can run concurrently and their access to common objects is serialized by locking.
3. Each sub-transaction may fail independently of its parent and of the other transaction.
4. If a sub-transaction aborts, the parent transaction can sometimes chose an alternative sub-transaction to complete its task.
5. If one or more of the sub-transaction fails, the parent transaction could record the fact and then commit with the result that all the successful child transactions commit.

Rules for nested transactions are:

1. A transaction may commit or about only after its child transactions have completed.
2. When a sub-transaction completes, it makes an independent decision either to commit provisionally or to about and its decision to abort is final.

3. When a parent abort, all its childs are aborted.

4. When a child abort, its parent may decide whether to abort or not.

5. If the top-level transaction commits, then all of the sub-transaction that have provisionally committed can commit too, provided that none of their ancestors are aborted.

Advantages

1. All the sub-transactions in the same level of hierarchy may run concurrently.

2. Sub-transactions may commit or abort independently.

Disadvantages

1. When a parent abort, all its subtransactions are aborted.

2. A transaction may commit or abort only after its child transactions have completed.

3. When a sub-transaction aborts, the parent can decide whether to abort or not.

10.5 Locks

Each server maintains locks for its own objects. Locks can not be released until the transaction has been committed or aborted at all servers. Distributed deadlocks might occur if different servers impose different ordering on transactions. This is generally used to maintain data consistency and serializability. There are two general types of locks. They are:

• **Shared:** if the transaction has locked the data object in shared mode, other transactions can concurrently lock it but only in shared mode. As we know the pairs of read operation from different transactions do not conflict, an attempt to set a read lock on an object with a read lock is always successful. All the transactions reading the same object can share their read lock.

• **Exclusive:** if a transaction has locked the data object in exclusive mode, no other transaction can lock it in any mode. In this locking scheme the server attempts to lock any object that is about to be used by any operation of a client transaction. If a

Transaction T_1		Transaction T_2	
Operations	**Locks**	**Operations**	**Locks**
Open transaction			
Bal.=b.get bal. ( )	Lock B	Open transaction	
b. set bal. (bal. *101)		Bal.=b.get bal. ( )	Wait for T_1
a. withdraw (bal./10)	Lock A	lock	
			For B
Close transaction	Unlock A,B		Lock B
		b. set bal. (bal.* 1.1)	
		c. withdraw (bal.10)	Lock C
		close transaction	Unlock B,C

Fig. 10.5: Transaction T_1 and T_2 with exclusive locks

client request access to an object that is already locked due to another transaction, the request is suspended and the client must wait until the object is unblocked (Fig. 10.5).

Types of locking (locking algorithm)

In lock-based concurrency control algorithms, a transaction must lock a data object before accessing it. A transaction is well performed if it:

- Locks the data object before accessing it
- Does not lock the data object more than once
- Unlock all the data objects before it completes.

Types of locking algorithms are:

   i. Static locking
  ii. Two-phase locking.
 iii. Strict two-phase locking.

i. Static locking: In this, a transaction acquires locks on all data objects it needs before executing any action on data objects. Static locking requires a transaction to pre-declare all the data objects it needs for execution. A transaction unlocks all the locked data objects only after it has executed all its actions.

Advantages

- It is conceptually simple.
- It is easy to implement.

Disadvantages

- It requires prior knowledge of data.
- It limits the concurrency.

ii. Two-phase locking: It is a dynamic locking scheme in which a transaction requests a lock on a data object when it needs them. It ensures the conflict-serializable schedules. Every transaction executes in two phases. The two phases can be explained as:

Phase 1: Growing phase

  a. Transaction may obtain locks
  b. Transaction may not release lock

Phase 2: Shrinking phase

  a. Transaction may release locks
  b. Transaction may not obtain lock

Advantages

- It maintains database concurrency.
- It increases concurrency over static locking.

Disadvantages

- Deadlock
- Cascaded aborts/rollbacks.

iii. **Strict two-phase locking:** To prevent from dirty reads and premature writes, strict executions are required and also to avoid cascading rollback which generally occurs under two-phase locking. So, enforce all this, any locks are applied during the progress of transaction are held until the transaction either commits or aborts. This is called strict-two-phase locking. It is beneficial although it reduces concurrency as a transaction holds locks for a longer period than required for consistency.

10. 6 Optimistic Concurrency Control

This approach was given by Kung and Robinson to avoid major drawbacks of locking scheme, as it is based on the observation that in most applications the likelihood of two clients transaction accessing the same object is low. In this approach each and every transaction goes through three phases:

  i. Working phase
  ii. Validation phase
  iii. Update phase.

10.6.1 Working phase

During this phase, each transaction has a tentative version of each of the object that it updates. This allows the transaction to abort either during the working phase or other validation phase. The rules for read/write are:

a. Read operations are performed if the tentative version for that transaction already exists.

b. Write operation record the new values of several concurrent transactions.

10.6.2 Validation phase

When the close transaction request is received, the transaction is validated to establish whether or not its operations on objects conflicts with operations of other transaction on same objects.

10.6.3 Update phase

If the transaction is validated, all the changes recorded in its versions are made permanent, the read only transactions can commit immediately after passing validation phase.

10.7 Timestamp Ordering

Each transaction is assigned a unique timestamp value when it starts and every transaction is validated when it is carried out and if operation cannot be validated, then the transaction is aborted immediately. The timestamp assigned ($ts(T_i)$) is globally unique which is ensured by using Lampert's algorithm. Every transaction is assigned a read timestamp (rts) and write timestamp (wts), such that:

  rts (x) : Largest timestamp of read on x.

  wts (x) : Largest timestamp of write on x.

Example: If T wants to read (X) :

  a. if $TS(T) < WTS(X)$ then read is rejected, T has to abort.

  b. Else read is accepted $RTS(X)$ updated.

Example: if T wants to write (X) :

  a. if $TS(T) < RTS(X)$ then write is rejected, T has to abort.

  b. if $TS(T) < WTS(X)$ then write is rejected, T has to abort.

  c. Else, allow the write, and update $WTS(X)$.

10.7.1 Multiversion timestamp ordering

This kind of schemes keeps old versions of data item to increase concurrency and each successful write creates the new version of data item.

Method

Each data item Q has a sequence of versions $<Q_1, Q_2, ..., Q_m>$ each version QK contains three data fields. They are:

  i. Content: The value of version QK.

  ii. W-timestamp (QK): Timestamp of the transaction that created (wrote) version QK.

  iii. R-timestamp (QK): Largest timestamp of a transaction that successfully reads version QK.

When a timestamp T_i creates a new version QK of q, QK's W-timestamp and R-timestamp are initialized to $TS(T_i)$. R-timestamp of QK is updated, whenever a transaction T_j reads QK, and $TS(Tj) > R$ - timestamp (QK).

10.8 Comparison of Methods for Concurrency Control

Two-phase locking	Optimistic methods	Timestamp ordering
It decides the serialization order dynamically	Allow transaction to proceed without any form of conflicts checking	It decides the serialization order statically
When conflict access to an object is detected, it makes transaction wait	When conflict access to an object is detected all transactions are allowed to proceed but some are aborted when they attempt to commit	When conflict access to an object is detected, it aborts the transaction immediately
Better than time stamp ordering for update-dominated transaction	A substantial amount of work may have to be repeated when a transaction is aborted	Better than locking for read-dominated transactions
Pessimistic methods	Optimistic methods	Pessimistic methods

SUMMARY

A transaction is a unit of program execution that access and possibly updates various data items. A transaction must see a consistent database. During transaction execution database may be, inconsistent. When the transaction is committed, the database must be consistent. A transaction is an agreement, communication or movement carried out between separate entities or objects, often involving the exchange of items of value such as information. To provide reliable units of work that allows correct recovery from failures and keep a database consistent even in case of failure. To provide isolation between programs accessing a database concurrently. Without isolation the program's outcomes are possibly erroneous, to preserve integrity of data, the database system must ensure:

Atomicity: Either all operations of the transaction are properly reflected in the database or none.

Consistency: Execution of a transaction in isolation preserves the consistency of the database isolation: Although multiple transactions may execute concurrently, each transaction must be unaware of other concurrently executing transactions. Intermediate transaction results must be hidden from other concurrently executed transactions.

Durability: After a transaction completed successfully, the changes it has made to the database persist, even if there are system failures.

Nested transactions are implemented differently in databases. In this the outermost transaction in the set is known as top-level transaction and all other transactions are known as sub-transactions.

REVIEW QUESTIONS

1. What do you understand by optimistic concurrency control?
2. Explain transaction and its ACID properties.
3. Define flat and nested transactions.
4. Write a short note on timestamp multi-version ordering.
5. Give the comparison methods for concurrency control.
6. What are locks and how can they be achieved?

11

Distributed Transaction

Contents

11.1 Introduction

It refers to those transactions in which objects are managed by multiple servers. When a distributed transaction comes to an end, the atomicity property of transactions requires either all of the services involved commit the transaction or all of them abort the transaction (Fig. 11.1). To achieve this, one of the servers takes on a coordinator's role, which involves ensuring the same outcome at all the servers. The manner in which the coordinator achieves this depends on the protocol chosen. A protocol allows servers to communicate with one another to reach a joint decision as to whether to commit or abort. Distributed transactions must be serialized globally. This can be achieved with the help of locks that helps in maintaining the serializability of the transaction.

Distributed transactions are hard because they strictly follow the ACID properties of transaction, which are:

• Atomic: Ensuring all are committed in case of distributed transaction.

• Consistent: Failure may affect only a part of transaction.

• Isolated: Commitment must occur "simultaneously" at all sites.

• Durable: Not much different when other problems solved. It also makes "delayed commit" difficult.

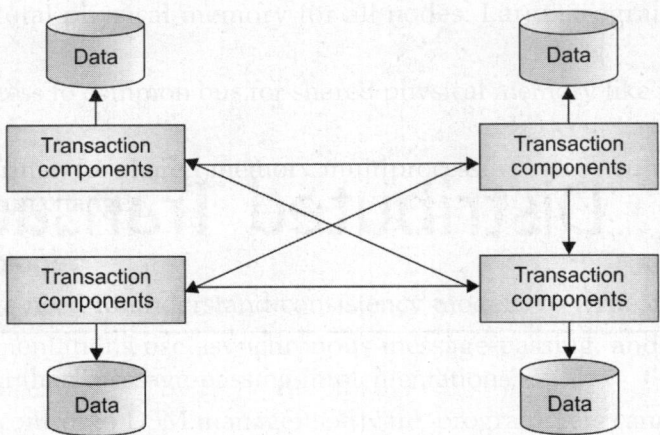

Fig.11.1: Distributed database transactions

11.2 Flat and Nested Distributed Transaction

A client transaction becomes distributed if it invokes operations on several different servers. There are two different ways that distributed transactions can be structured:

i. Flat transaction: In this client makes request to more than one server. Transaction T is a flat transaction that invokes operations on objects in servers X, Y and Z. Transactions completes each of its requests before going on to the next one. In every transaction client can access the objects sequentially. If server uses locking, a transaction will only be waiting for one object at a time (Fig. 11.2a).

ii. Nested transaction: Top-level transaction can open sub-transaction and each sub-transaction can open further sub-transaction down to any depth of nesting. Client's transaction T that opens two sub-transactions T_1 and T_2, which access objects at servers X and Y. They open their sub-transactions further as T_{11}, T_{12}, T_{21} and T_{22} on server M, N, O and P respectively. The transactions at same level can run concurrently and the transactions at different servers can run in parallel (Fig. 11.2b).

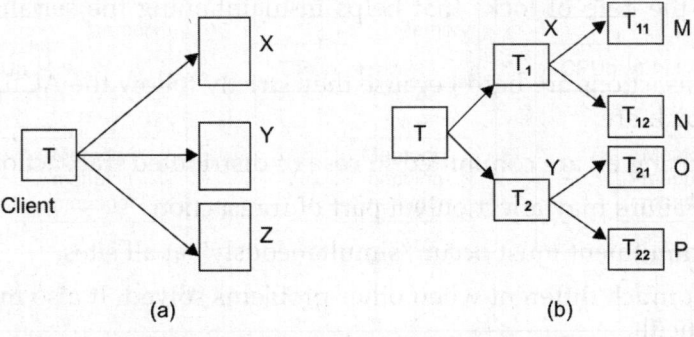

Fig 11.2: (a) Flat transaction; (b) Nested transaction

11.3 Atomic Commit Protocols

1. Transaction atomicity requires at the end
   - either all of its operations are carried out or none of them
2. In a distributed transaction, the client has requested the operations at more than one server
3. One-phase atomic commit protocol
   - The coordinator tells the participants whether to commit or abort
   - This does not allow one of the servers to decide to abort—it may have discovered a deadlock or it may have crashed and been restarted
4. Two-phase atomic commit protocol
   - It is designed to allow any participant to choose to abort a transaction
   - Phase 1: Each participant votes. If it votes to commit, it is prepared. It cannot change its mind. In case it crashes, it must save updates in permanent store
   - Phase 2: The participants carry out the joint decision
5. The decision could be commit or abort participants record it in permanent store.

11.3.1 Two phase commit protocols

During the progress of a transaction, the only communication between coordinator and participant is the joint request. The client requests to commit or abort delivers to the coordinator

- If client or participant requests abort, the coordinator informs the participants immediately
- If the client asks to commit, the 2PC comes into use.

Two-phase commit (2PC)

i. Voting phase: coordinator asks all participants, if they can commit
   - if yes, participant records updates in permanent storage and then votes.
ii. Completion phase: coordinator tells all participants to commit or abort.

Participant interface

- canCommit?(trans) → Yes/No
  - Call from coordinator to participant to ask whether it can commit a transaction. Participant replies with its vote
- doCommit(trans)
  - Call from coordinator to participant, to tell participant to commit its part of a transaction
- doAbort(trans)
  - Call from coordinator to participant to tell participant to abort its part of a transaction.

Coordinator interface

- haveCommitted(trans, participant)
  - Call from participant to coordinator to confirm that it has committed the transaction
- getDecision(trans) → Yes / No
  - Call from participant to coordinator to ask for the decision on a transaction after it has voted 'Yes' but has still no reply after some delay. Used to recover from server crash or delayed messages.

Phase 1 (voting phase):

1. The coordinator sends a canCommit? Request to each of the participants in the transaction.
2. When a participant receives a canCommit? Request it, replies with its vote (Yes or No) to the coordinator. Before voting Yes, it prepares to commit by saving objects in permanent storage. If the vote is No the participant aborts immediately.

Phase 2 (completion according to outcome of vote):

3. The coordinator collects the votes (including its own).
   a. If there are no failures and all the votes are Yes, the coordinator decides to commit the transaction and sends a doCommit request to each of the participants.
   b. Otherwise the coordinator decides to abort the transaction and sends doAbort requests to all participants that voted Yes.
4. Participants that voted Yes are waiting for a doCommit or doAbort request from the coordinator. When a participant receives one of these messages it acts accordingly and in the case of commit, makes a haveCommitted call as confirmation to the coordinator.

Time-out actions in 2PC:

- To avoid blocking forever when a process crashes or a message is lost
  i. Uncertain participant (step 2) has voted yes. It cannot decide on its own
- It uses getDecision method to ask coordinator about the outcome
  i. Participant has carried out client requests, but has not had a commit from the coordinator. It can abort unilaterally
  ii. Coordinator delayed in waiting for votes (step 1). It can abort and send doAbort to participants.

11. 4 Concurrency Control in Distributed Transaction

Set of objects is managed by each server and is also responsible for existing that they remain consistent when they accessed by concurrent transactions. Each server is responsible for applying concurrency control to its own objects.

11.4.1 Locking

- Locks are used to order transactions according to the order of arrival of their operations at the same data item. The locks on an object are held locally.

- The local lock manager can decide whether to grant a lock or make the requesting transaction wait. Resource locking in a distributed system can be implemented with a lock manager. One process serves as a lock manager.
- Processes request a lock for a resource and then they either are granted a lock or wait for it to be granted when another process releases it.
- As lock managers at different servers set their locks independently of one another, it is possible servers may impose different orderings on transactions.

11.4.2 Timestamp ordering concurrency control

- Timestamps are used to order transactions according to their starting times. In distributed transactions, we require that each coordinator issue globally unique timestamps.
- Globally unique timestamp is issued to the client by the first coordinator accessed by a transaction. The transaction timestamp is passed to the coordinator at each server whose object performs an operation in the transaction.
- In distributed transaction, all servers are jointly responsible for ensuring the operation to be in serially equivalent manner. Even if local clock is not synchronized, then also same ordering of transactions can be achieved.
- The basic timestamp ordering rule: A transaction's request to write an object is valid only if the object was last read and written by earlier transactions. A transaction's request to read an object is valid only if that object was last written by earlier transaction.
- In this method, conflicts are resolved as each operation is performed. If the resolution of a conflict requires a transaction to be aborted, the coordinator will be informed and it will abort the transaction of all the participants.

11.4.3 Optimistic concurrency control

- If we assume that, validation is fast in a single server. But in distributed transaction of two phase commit protocol, it takes some times and it delays other transactions from entering validation until a decision on the current has been obtained.
- In distributed optimistic concurrency control, each server applies a parallel validation protocol. If parallel validation is used, transaction will not suffer from commitment deadlock.
- If a server simply performs independent validation, it is possible that different servers of a distributed transaction may serialize the same set of transaction in different orders.
- All of the servers of a particular transaction use the same globally unique transaction number at the start of the validation.

11.5 Distributed Deadlock

11.5.1 Centralized deadlock detection

In this each server sends its local wait-for-graph and the central deadlock detector checks a cycle by global wait-for-graphs. Phantom deadlocks happen when one of the transactions that hold a lock (and creates deadlock) will have aborted during deadlock detection phase.

11.5.2 Distributed deadlock detection

The algorithms required to detect deadlock in the distributed environment are called edge chasing or path pushing. There is no requirement of global wait-for-graph. The mechanism for detecting deadlock in distributed environment can be explained as:

1. The lock manager informs the coordinator when transactions start waiting and when they become active again.
2. The three phases are initiated:
    i. If transaction A starts waiting for transaction B waiting to access a data item at another server, transaction B's server sends a probe containing the wait-for-relationship to the server of data items where transaction B is blocked and all the servers in which transactions share lock with transaction B.
    ii. If the data item is hold by another transaction (by consulting with coordinator), add this relationship to the probe and forward the probe in the same manner as above resolution.
    iii. When cycle is detected, a transaction in a cycle is aborted to break the deadlock.

11.6 Transaction Recovery

1. Transaction recovery is required to ensure failure atomicity and durability in the presence of failure
    – Transaction abort: Update-in-place and deferred update.
    – System crash: Logging and shadow versions.
2. A transaction processing service is 'fault tolerant' if it fails temporarily but can recover without loss of data.
3. There are many variants. We are concerned with fault tolerance in a simple distributed transaction processing service.
4. Atomicity means all or nothing; a requirement that
    – the effects of all committed transactions are reflected in the data items and
    – none of the effects of incomplete or aborted transactions are reflected in the data items.
5. Durability and failure atomicity
    • Durability ensures that data items are saved in permanent storage and therefore when a two-phase protocol commits, the changes to the data item are permanently stored.
    • Failure atomicity ensures that the effects of a transaction are atomic even when a client or server fails.
6. The recovery file is read backwards, restoring committed data items as necessary.
    • This is made more effective by the use of the intention lists.
    • Checkpointing prevents the necessity to read the whole recovery file on failure.

11.6.1 Recovery schemes

1. Logging
2. Checkpointing
3. Shadow versions

1. Logging

i. This is one of the uses of the recovery file (in this instance it is sometimes called the log).

ii. It is a log of all the values of data items, transaction status entries and intention lists of transactions processed by the server.

iii. Normally the recovery manager is called whenever a transaction prepares to commit, commits or aborts a transaction.

iv. When a server is prepared to commit a transaction, the RM appends to the recovery file.

v. All the data items in its intention list, followed by the current transaction status (prepared).

vi. When the transaction is committed or aborted, the RM appends the corresponding status of the transaction to the recovery file.

vii. Recovery process

   a. When a server is restarted, sets data to their default values and lets recovery manager proceed recovery

   b. Recovery manager reads recovery file "backward"

     • if transaction is committed, restore committed value

     • if not, record status of transaction as aborted

2. Checkpointing

i. Process of reclaiming space in recovery file

   a. To reduce space required for logging

   b. To improve recovery process

ii. Record in recovery file can be discarded except

   a. Current committed value of data items

   b. Transaction status entries and intention lists of transactions that have not yet been fully resolved

iii. Checkpoint

Mark in recovery file where checkpoint process has begun.

3. Shadow versions

i. The previous recovery technique stored all the recovery data in the log file that is read in reverse during the recovery process.

ii. The shadow versions technique reduces the details stored in the recovery file and uses a map to access data items held in a version store.

iii. When a transaction prepares to commit, changed data items are appended to the version store.

iv. The versions written are shadows of previous committed versions.

v. When a transaction commits, a new map is made. The switch between the two maps must be done atomically.

vi. To recover, the recovery manager reads the map file to access the data items.

SUMMARY

A distributed transaction is an operation bundle, in which two or more network hosts are involved. Usually, hosts provide transactional resources, while the transaction manager is responsible for creating and managing a global transaction that encompasses all operations against such resources. Distributed transactions, as any other transactions, must have all four ACID properties, where atomicity guarantees all-or-nothing outcomes for the unit of work (operations bundle). A common algorithm for ensuring correct completion of a distributed transaction is the two-phase commit (2PC). The algorithm is usually applied for updates able to commit in short period of time, ranging from couple of milliseconds to couple of minutes. Flat transactions represent the simplest form of transaction, and for almost all existing system, it is the only one that is supported at the application programming level. A flat transaction is the basic building block for organizing an application. The transactions must maintain the property of consistency and durability.

REVIEW QUESTIONS

1. What do you understand by distributed transaction?
2. Describe fault tolerance services in distributed transactions.
3. Write the differences between flat and nested distributed transaction.
4. Explain atomic commit protocols.
5. Write short notes on:
    - i. Transaction recovery
    - ii. Concurrency control in distributed transaction
    - iii. Replication

Replication

Contents

12.1 Introduction

Replication refers to the maintenance of copies at multiple sites. Replication is a technique for enhancing services. A logical object is implemented by a collection of physical copies called replicas. It is a key for providing high availability and fault tolerance in distributed systems, hence used widely. It is a technique for enhancing services by maintaining the local copies. The motivation is to improve a service's performance, to increase its availability or to make it fault tolerant.

1. **Performance enhancement:** the caching of data at clients and servers is by now familiar as a means of performance enhancement.

2. **Fault tolerance:** highly available data is not necessarily strictly correct data, guarantees strictly correct behavior despite a certain number and type of faults. It requires strict data consistency (if a process reads any memory location), the value returned by the read operation is the value written by the most recent write operation to the location between all replicated servers. Replication of read-only data is simple, but replication of mutable data incurs overheads in form of the protocol.

3. **Increase availability:** Factors that affect availability are server failures and network partitions. User requires services to be highly available. The availability of the service is that, if there are n replicated servers each of which would crash in a probability of p.

12.2 System Model and Group Communication

12.2.1 System model

The data consist of a collection of terms called objects and each object is implemented by a collection of physical copies called replicas. The system model provides managing replicas and describes group communication.

The models involve replica cells by distinct replica managers which are the components that contain the replicas on a given computer and perform operations upon them directly.

A full implementation of group communication incorporates a group membership service to manage the dynamic membership of groups in addition to multicast communication (Fig. 12.1). Multicast and group membership management are strongly inter-related. The group communication service has to manage chance in the group's membership while casts take place concurrently. A group membership service has four main tasks, they are:

1. Providing an interface for group membership changes
2. Implementing a failure detector
3. Notifying members of group membership changes
4. Performing group address expansion

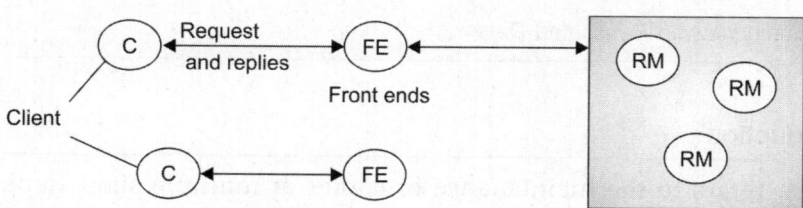

Fig. 12.1: Architecture mode for the management of replicated data

12.2.2 Group communication

It occurs when a message is sent to the group of processes not to the single process. They are generally used in:

1. *Fault tolerant based on replicated services*: The service is implemented by a group of server processes. A client process multicast request message to the server group. Each group member performs identical operation on a request.
2. *Locating objects in distributed services*: In this files are not replicated and hence the request message is sent to the server group. The member holding the requested file responds.
3. *Updating replicated Data*: Data may be replicated to improve performance and reliability. Update request is multicasted to the server group.
4. *Multiple notification*: A group needs to be notified of certain events.

A full group membership service maintains group views which are lists of the current group members identified by their unique process identifiers, the list ordered according to the sequence in which the members joined group. A new group view is generated when processes are added or executed. A false suspicion of process and consequent exclusion of the process from group may reduce group's effectiveness.

12.3 Faults Tolerant Service

For fault tolerant service, we provide a service that is correct despite process failure by replicating data and functionality at replica managers. For the sake of simplicity we assume that communication remains reliable and that no possible error occurs. Each replica manager is assumed to behave according to the specification at the semantics of the object; it manages, when they have not crashed (Fig. 12.2).

Example: A bank account specification would include an assurance that fund transferred between accounts can never disappear and that only deposits and withdraws affect the balance at any particular account.

12.3.1 Passive replication

In passive model of replication for fault tolerance there is at any one time a single primary replica manager and one or more secondary replica managers backup or slaves. In pure form of model, front end commutate only with primary replica managers to obtain services. The primary replica managers execute operations and sends copies of updated data to backups. If primary fails, one of backup is promoted to act as the primary.

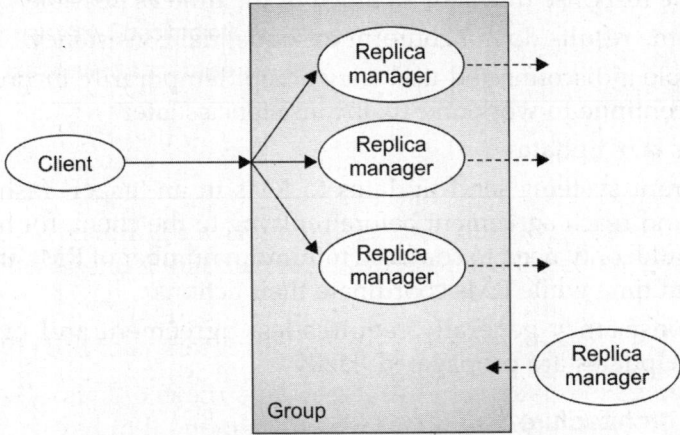

Fig. 12.2: Passive mode for fault tolerance

12.3.2 Active replication

In active mode of replication for fault tolerance the replica managers are state machines that play equivalent roles and are organized as a group. Front ends multicast their request to the group of replica managers and the entire replica manager process the request independently but identically and reply. If any replica manager crashes. Then this has no impact upon the performance of service since the remaining replica managers continue to responds in a normal way (Fig. 12.3).

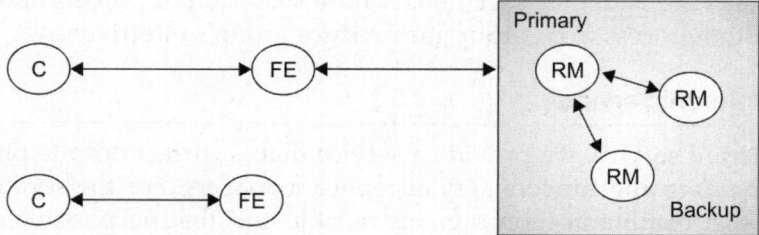

Fig. 12.3: Active replication

12.4 Highly Available Services

Replication technique is used to make services highly available, i.e. clients access the services with reasonable request time. Several systems that provide the highly available services are:

1. Gossip
2. Bayou
3. Coda

1. We discuss the application of replication techniques to make services highly available.
   - We aim to give clients access to the service with
   - Reasonable response times for as much of the time as possible
   - Even if some results do not conform to sequential consistency
   - For example a disconnected user may accept temporarily inconsistent results if they can continue to work and fix inconsistencies later

2. Eager versus lazy updates
   - Fault tolerant systems send updates to RMs in an 'eager' fashion (as soon as possible) and reach agreement before replying to the client, for high availability, clients should: only need to contact a minimum number of RMs and be tied up for a minimum time while RMs coordinate their actions
   - Weaker consistency generally requires less agreement and makes data more available. Updates are propagated 'lazily'

12.4.1 Gossip architecture

1. The gossip architecture is a framework for implementing highly available services (Fig. 12.4)
   i. Data is replicated close to the location of clients
   ii. RMs periodically exchange 'gossip' messages containing updates

2. Gossip service provides two types of operations
   i. Queries—read only operations
   ii. Updates—modify (but do not read) the state

3. FE sends queries and updates to any chosen RM

   One that is available and gives reasonable response times.

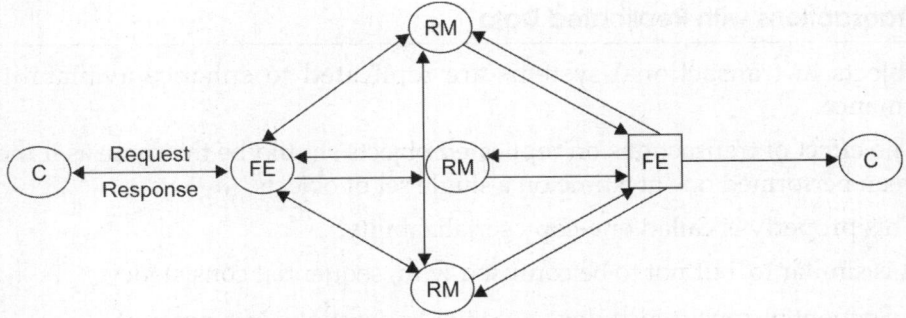

Fig.12.4: Query and update operation in a gossip service

4. Two guarantees (even if RMs are temporarily unable to communicate:

   i. Each client gets a consistent service over time (i.e. data reflects the updates seen by client, even if the use different RMs). Vector timestamps are used— with one entry per RM.

   ii. Each relaxed consistency between replicas. All RMs eventually receive all updates. RMs use ordering guarantees to suit the needs of the application (generally causal ordering). Client may observe stale data.

5. The service consists of a collection of RMs that exchange gossip messages.

6. Queries and updates are sent by a client via an FE to an RM.

7. The six phases in executing a client request are:

   i. Request
   - FEs normally use the same RM and may be blocked on queries
   - Update operations return to the client as soon as the operation is passed to the FE

   ii. Update response
   The RM replies as soon as it has seen the update

   iii. Coordination
   - The RM waits to apply the request until the ordering constraints apply.
   - This may involve receiving updates from other RMs in gossip messages

   iv. Execution
   The RM executes the request

   v. Query response
   If the request is a query, the RM now replies

   vi. Agreement
   RMs update one another by exchanging gossip messages (lazily), e.g. when several updates have been collected or when an RM discovers it is missing an update.

12.5 Transactions with Replicated Data

The objects in transactional systems are replicated to enhance availability and performance:

i. The effect of transactions on replicated objects should be the same as if they had been performed one at a time on a single set of objects.

ii. This property is called one-copy serializability.

iii. It is similar to, but not to be confused with, sequential consistency.

- Sequential consistency does not take transactions into account

iv. Each RM provides concurrency control and recovery of its own objects

- We assume two-phase locking in this section

v. replication makes recovery more complicated

- When an RM recovers, it restores its objects with information from other RMs.

12.5.1 Architectures of replicated transactions

1. We assume that an FE sends requests to one of a group of RMs. In the primary copy approach, all FEs communicate with a single RM which propagates updates to back-ups.

2. In other schemes, FEs may communicate with any RM and coordination between RMs is more complex. An RM that receives a request is responsible for getting cooperation from the other RMs rules as to how many RMs are involved, vary with the replication scheme

- For example in the read one/write all scheme, one RM is required for a read request and all RMs for a write request.

3. It propagates requests immediately or at the end of a transaction. In the primary copy scheme, we can wait until end of transaction (concurrency control is applied at the primary)– but if transactions access the same objects at different RMs, we need to propagate the requests so that concurrency control can be applied

4. Two-phase commit protocol: becomes a two-level nested 2PC. If a coordinator or worker is an RM, it will communicate with other RMs that it passed requests during the transaction.

12.5.2 Available copies replication

1. The simple read one/write all scheme is not realistic

- Because it cannot be carried out if some of the RMs are unavailable,
- Either because they have crashed or because of a communication failure

2. The available copies replication scheme is designed to allow some RMs to be temporarily unavailable

A read request can be performed by any available RM—writes requests are performed by the receiving RM and all other available RMs in the group.

Available copies

1. Local concurrency control achieves one-copy serializability provided the set of RMs does not change. But we have RMs failing and recovering.

2. An RM can fail by crashing and is replaced by a new process. The new process restores its state from its recovery file.

3. FEs use timeouts in case an RM fails then try the request at another RM in the case of a write, the RM passing it on may observe failures.

4. If an RM is doing recovery, it rejects requests (& FE tries another RM): For one-copy serializability, failures and recoveries are serialized with respect to transactions that is, if a transaction observes that a failure occurs, it must be observed before it started or after it finished.

5. One-copy serializability is not achieved if different transactions make conflicting failure observations.

Quorum consensus methods

1. To prevent transactions in different partitions from producing inconsistent results make a rule that operations can be performed in only one of the partitions. RMs in different partitions cannot communicate each subgroup decides independently whether they can perform operations.

2. A quorum is a subgroup of RMs whose size gives it the right to perform operations, e.g. if having the majority of the RMs could be the criterion in quorum consensus schemes.

3. Update operations may be performed by a subset of the RMs and the other RMs have out-of-date copies version numbers or timestamps are used to determine which copies are up-to-date operations and are applied only to copies with the current version number.

SUMMARY

To achieve high availability and fault tolerance the concept of replication is used in distributed transaction. In active replication all replica manager processes request independently. In primary backup replication, fault tolerance is achieved by directly sending all request through a distinguish replica manager and have backup replica manager to take over if this fails. Group communication is in both replications. Replica manager exchange updates with one another when they become reconnected. Replication takes into account for the possibility of partition. Each replica manager is assumed to behave according to the specification of the semantics of the object it manages. Replication scheme are designed with the assumption that partitions will eventually be repaired.

REVIEW QUESTIONS

1. Write a short notes on replication and also explain fully the replicated database.

2. What do you understand by highly available services?

3. Explain gossip architectures.

4. Discuss system models and the concept of group communication.

5. Define transaction with replicated database.

6. What do you mean by active replication?

MODEL QUESTION PAPER – I
B Tech
THEORY EXAMINATION (VII SEMESTER)
DISTRIBUTED SYSTEM

Total Marks: 100

Time: 3 Hours

Instructions: i. All questions are compulsory.
ii. All questions carry equal marks.

SECTION A

(Answer in thirty words only)   2 × 10 = 20

Note: All parts are compulsory

   A. Define Lamport's logical clock.

   B. Explain consensus agreement problem.

   C. What do you understand by flat transaction.

   D. Define mounting and hints in DFS.

   E. Brief the limitation of distributed system.

   F. Explain Domino effect.

   G. Explain system model for distributed deadlock.

   H. Give the design issues of DSM.

   I. Explain forward recovery approach.

   J. Define token and non-token based algorithm for DS.

SECTION B

(Answer in approximately 250 words)   10 × 5 = 50

Note: Answer any **FIVE** of the following

A. Discuss vector clock and also give its advantages over logical clock. Explain the implementation rules of vector clock. Give the vector timestamp of messages for following schematic, where S_1, S_2 and S_3 represent the processing sites.

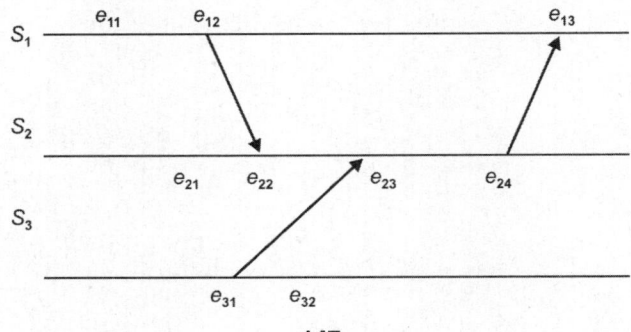

B. Explain Ricart–Agarwala's algorithm and Shinghal's heuristic algorithm for mutual exclusion. How many messages per critical section execution are required?

C. Explain all the algorithms for implementing distributed shared memory.

D. Explain 2PL and strict 2PL in detail.

E. With reference to the backward recovery approach explain the operation based approach and state based approach.

F. How will you execute the transaction when data is replicated among several sites and data consistency is highest criteria? Also discuss the case when data availability is the highest criteria.

G. Discuss causal ordering of messages. Give all algorithms which can order the messages according to causal dependencies.

H. Explain two phase and three phase commit protocols in detail.

SECTION C

(Answer in approximately 400 words)   **15 × 2=30**

Note: Answer any **TWO** of the following:

A. i. Show that Byzantine agreement can not always be reached among four processors if two processors are faulty.

  ii. Show that a solution to the consensus problem can be used to solve the interactive consistency problem.

B. i. Comment on fault tolerance services in distributed system.

  ii. Discuss the optimistic method for distributed concurrency control. What are the different validation conditions for optimistic concurrency control?

C. Explain in context of distributed system:

  i. Deadlock detection and prevention

  ii. Obermark's path pushing algorithm

  iii. Atomic commit in distributed database

MODEL QUESTION PAPER – II
B Tech
THEORY EXAMINATION (VII SEMESTER)
DISTRIBUTED SYSTEM

Total Marks: 100

Time: 3 Hours

Instructions: i. All questions are compulsory.

ii. All questions carry equal marks.

SECTION A

(Answer in thirty words only)   2 × 10 = 20

Note: All parts are compulsory

A. Define distributed system.

B. Explain interactive consistency agreement problem.

C. Define token and nontoken based algorithm for DS.

D. Define caching and bulk data transfer in DFS.

E. Brief the limitation of distributed system.

F. Explain Domino effect.

G. Explain fundamental model for distributed system.

H. Give the design issues of DSM.

I. Explain forward recovery approach.

J. What do you understand by nested transaction.

SECTION B

(Answer in approximately 250 words)   10 × 5 = 50

Note: Answer any **FIVE** of the following

A. Discuss vector clock and also give its advantages over logical clock. Explain the implementation rules of vector clock. Also quote an example.

B. Explain Lamport's algorithm and Raymond's tree algorithm for mutual exclusion.

C. Explain all the algorithms for implementing distributed shared memory.

D. Explain the Concept of global state with example.

E. With reference to the backward recovery approach. Explain the operation based approach and state based approach.

F. How will you execute the transaction when data is replicated among several sites and data consistency is the highest criteria? Also discuss the case when data availability is the highest criteria.

G. Write a short note on:
   i. Voting protocol
   ii. Dynamic voting protocol.
H. Explain two phase and three phase commit protocols in detail.

SECTION C

(Answer in approximately 400 words)   15 × 2 = 30

Note: Answer any **TWO** of the following

A.   i. Give the OM algorithm for byzantine agreement problem.

   ii. Give the performance metrics for mutual exclusion algorithms.

   iii. Give solutions to impossible result for byzantine agreement problem.

B.   i. Give the algorithm for termination detection.

   ii. Explain timestamp ordering.

   iii. Define transaction with replicated data.

C. Explain in context of distributed system:

   i. The Ho-Ramamurthy algorithm.

   ii. Edge chasing algorithm.

   iii. Concurrency control in distributed transaction.

UPTU QUESTION PAPER – I
B Tech
THEORY EXAMINATION 2012–13 (VII SEMESTER)
DISTRIBUTED SYSTEM

Total Marks: 100

Time: 3 Hours

Instructions: i. All questions are compulsory.

ii. All questions carry equal marks.

1. Attempt any two parts of the following   (10 × 2 = 20)

a. Discuss the relative advantages and disadvantages of the various commonly used models for configuring distributed computing systems.

b. Discuss the major issue in designing a distributed system.

c. How Lamport's clock casually relate two events? Discuss the limitations of Lamport's clock. How the vector clocks remove the limitations of Lamport clock? Explain.

2. Attempt any two parts of the following: (10 × 2 = 20)

a. What is deadlock? What are the necessary conditions for the occurrence of deadlock in distributed system? Describe the deadlock handling strategies in distributed system.

b. Classify the deadlock detection algorithms. Describe the path pushing deadlock detection algorithm.

c. Write and explain a token based algorithm for mutual exclusion. Describe its performance on important metrics.

3. Attempt any two of the following: (10 × 2 = 20)

a. Describe Byzantine agreement problem, and explain its solution. Show that Byzantine agreement cannot always be reached among four processors if two processors are faulty.

b. Describe mechanism for building distributed file system. Explain data access actions in distributed file system.

c. Discuss the architecture of distributed shared memory and its advantages.

4. Attempt any two of the following: (10 × 2 = 20)

a. What is livelock problem in message passing system? How the synchronous checkpointing methods avoid the livelock problem? Describe.

b. Describe two phase commit protocol. How the protocol handles the site failure? Write and explain its limitations.

c. What do you understand by dynamic voting? Explain dynamic voting protocol in brief.

5. Write short notes on any two of the following: (10 × 2 = 20)

a. i. Briefly explain the objectives of distributed transaction management.

   ii. What is lock? Describe the functions of lock manager.

b. i. Describe how a non-recoverable situation could arise if write locks are released after the last operation of a transaction but before its commitment.

   ii. Draw a schematic diagram of the distributed transaction management model. Explain each component in brief.

c. i. Define and differentiate the simple and nested distributed transactions.

   ii. What is atomic commit protocol? Explain in brief.

UPTU QUESTION PAPER – II
B Tech
THEORY EXAMINATION 2013–14 (VII SEMESTER)
DISTRIBUTED SYSTEM

Total Marks: 100

Time: 3 Hours

Instructions: i. All questions are compulsory.

ii. All questions carry equal marks.

1. Attempt any four of the following: (5 × 4 = 20)

  a. How the distributed computing system is better than parallel processing system? Explain.

  b. Discuss the impact of the absence of global clock in distributed systems.

  c. Define the term transparency. Explain important types of transparencies in distributed system.

  d. What is termination detection in distributed system? Explain any algorithm for termination detection.

  e. What is vector clock? How this maintains causal ordering? Explain.

  f. Explain the following distributed computing model:

    i. Mini computer model

    ii. Workstation model

    iii. Workstation server model

2. Attempt any two parts of the following: (10 × 2 = 20)

  a. What is mutual exclusion? Describe the requirements of mutual exclusion in distributed system. Is mutual exclusion problem more complex in distributed system than single computer system? Justify your answer.

  b. What do you mean by deadlock avoidance? Explain in brief. Describe edge-chasing deadlock detection algorithm.

  c. Write and explain a nontoken-based mutual exclusion algorithm. Describe its merit and demerits.

3. Attempt any two of the following: (10 × 2 = 20)

  a. Classify the agreement problems. Explain the applications of agreement algorithms.

  b. Write and explain various issues that must be addressing design and implementation of distributed file system.

  c. Describe memory coherence. Briefly explain the write invalidate and write update protocols.

4. **Attempt any two parts of the following:** (10 × 2 = 20)

   a. What is checkpointing in message passing system? Explain the recovery in message passing system using asynchronous checkpointing scheme.

   b. i. Define the livelocks. What is the difference between a deadlock and livelock?

   ii. Show that when checkpoints are taken after every $K(K > 1)$ message sent, the recovery mechanism suffers from the domino effect. Assume that a process takes a checkpoint immediately after sending the Kth message but doing anything else.

   c. Describe three phase commit protocol. How three phase commit protocol is different than two phase commit protocol?

5. **Write short notes on any four of the following:** (10 × 2 = 20)

   a. Describe the advantages and disadvantages of multiversion timestamp ordering over the ordinary timestamp ordering.

   b. Describe the optimistic concurrency control method. How this method avoids the drawbacks of locking? Explain.

   c. i. What is phantom deadlock? Describe the conditions for the occurrence of phantom deadlock.

   ii. Describe the architecture of replicated transactions.

UPTU QUESTION PAPER – III
B Tech
THEORY EXAMINATION 2014–15 (VII SEMESTER)
DISTRIBUTED SYSTEM

Total Marks: 100

Time: 3 Hours

Instructions: i. All questions are compulsory.

ii. All questions carry equal marks.

1. Attempt any four of the following: (5 × 4 = 20)

a. What is a distributed system? Describe the main characteristics of distributed systems. Give two examples of distributed system.

b. What are commit protocols? Explain how two phase protocols respond to failure of participating site and failure of coordinator.

c. What do you mean by mutual exclusion in distributed system? What are requirements of a good mutual exclusion algorithm?

d. What are vector clocks? Explain with the help of implementation rule of vector clocks, how they are implemented? Give the advantages of vector clock over Lamport clock.

e. What is replication and replica manager? Give the architectural model for replicated data.

f. What is distributed shared memory (DSM). Explain with diagram the architecture of distributed shared memory.

2. Attempt any four of the following: (5 × 4 = 20)

a. Explain the following:

   i. Gossip architecture

   ii. Quorum consensus methods.

b. What do you mean by recovery in concurrent systems? Explain.

c. What is voting protocol? Explain static and dynamic voting protocols.

d. Explain the Ricart–Agrawala algorithm for mutual exclusion. Mention the performance of this algorithm.

e. Define fault and failure. What are different approaches to fault tolerance? Explain.

f. Describe the following algorithms for implementing DSM:

   i. Migration algorithm

   ii. Full-replication algorithm

3. Attempt any two parts of the following: (10 × 2 = 20)

  a. i. What are the goals of distributed transaction? Distinguish between flat and nested transaction along with its structure.

    ii. Explain optimistic concurrency control.

  b. Define forward recovery and backward recovery. List advantages and disadvantages of forward recovery. Explain two approaches of backward–error recovery.

  c. What are agreement protocols? Explain Byzantine agreement problem, consensus problem and interactive consistency problem. Describe Lamport–Shostak–Pease algorithm.

4. Attempt any two of the following: (10 × 2 = 20)

  a. What are the advantages and drawback of multiversion timestamp ordering in comparison to the basic timestamp ordering?

  b. Write short note on:

    i. Livelocks

    ii. Domino effects

    iii. Failure-resilient processes

    iv. Consistent Checkpoints.

  c. i. Explain typical architecture of distributed file system. Give the mechanisms for building distributed file system.

    ii. What is caching? How it is useful in DFS?

5. Write short notes on any two of the following: (10 × 2 = 20)

  a. Give the deadlock handling strategies in distributed systems? What are the differences in centralized, distributed and hierarchical control organizations for distributed deadlock detection?

  b. Why is scalability an important feature in the design of distributed system? Discuss some of the guiding principles for designing a scalable distributed system.

  c. Distinguish between:

    i. Resource deadlock and communication deadlock.

    ii. Token based and nontoken-based algorithm.

Index